Modern Cartography Series
MAPPING THE EPIDEMIC
A SYSTEMIC GEOGRAPHY OF COVID-19 IN ITALY

VOLUME 9

Modern Cartography Series
VOLUME 9

MAPPING THE EPIDEMIC
A SYSTEMIC GEOGRAPHY OF COVID-19 IN ITALY

Series Editor

D. R. FRASER TAYLOR

Edited by

EMANUELA CASTI
Emeritus professor of Geography, University of Bergamo, Italy

with

FULVIO ADOBATI
Professor of urban planning, University of Bergamo, Italy

ILIA NEGRI
Professor of statistics, University of Bergamo, Italy

Translated by

DAVIDE DEL BELLO

ELSEVIER

Elsevier
Radarweg 29, PO Box 211, 1000 AE Amsterdam, Netherlands
The Boulevard, Langford Lane, Kidlington, Oxford OX5 1GB, United Kingdom
50 Hampshire Street, 5th Floor, Cambridge, MA 02139, United States

Notices
Knowledge and best practice in this field are constantly changing. As new research and experience broaden our understanding, changes in research methods, professional practices, or medical treatment may become necessary.

Practitioners and researchers must always rely on their own experience and knowledge in evaluating and using any information, methods, compounds, or experiments described herein. In using such information or methods they should be mindful of their own safety and the safety of others, including parties for whom they have a professional responsibility.

To the fullest extent of the law, neither the Publisher nor the authors, contributors, or editors, assume any liability for any injury and/or damage to persons or property as a matter of products liability, negligence or otherwise, or from any use or operation of any methods, products, instructions, or ideas contained in the material herein.

Library of Congress Cataloging-in-Publication Data
A catalog record for this book is available from the Library of Congress

British Library Cataloguing-in-Publication Data
A catalogue record for this book is available from the British Library

ISBN: 978-0-323-91061-3
ISSN: 1363-0814

For information on all Elsevier publications
visit our website at https://www.elsevier.com/books-and-journals

Publisher: Candice Janco
Acquisitions Editor: Peter J. Llewellyn
Developmental Editor: Michelle Fisher
Production Project Manager: Bharatwaj Varatharajan
Cover Designer: Victoria Pearson

Typeset by STRAIVE, India

Contents

List of figures   ix
List of contributors   xi
Acronyms   xiii
Preface   xv
Editors' comments: The Covid-19 epidemic in Italy, a European epicenter   xvii

Introduction

Territorial analysis and reflexive mapping on the Covid-19 infection

Emanuela Casti

1 Context   1
2 Epidemiological data and the issue of sources   1
3 Stages of virus propagation in relation to social and territorial factors   3
4 Theoretical grounding and spatial dimension of the virus   6
5 Geographical implications of the contagion and Italian outcomes   8
6 From GIS to reflexive mapping   11
7 Conclusions   14
References   15

Chapter 1

Population and contagion spread

1.1 Evolution of epidemic outcomes in Europe

Elisa Consolandi

1.1.1 Premise   19
1.1.2 Covid-19 in Europe   20
1.1.3 Factors favoring viral propagation   24
1.1.4 Conclusions   27
References   27

1.2 Italy into three parts: The space–time spread of contagion

Emanuela Casti and Elisa Consolandi

1.2.1 Introduction   29
1.2.2 Mapping contagion: The spread of Covid-19 across the three Italies   30
1.2.3 Morpho-climatic and socio-territorial factors   34
1.2.4 Suggestions for a new territorial project   37
References   38
Further reading   39

1.3 Evolution and intensity of infection in Lombardy

Elisa Consolandi

1.3.1 Premise   41
1.3.2 The Lombard territory: Distribution and temporal evolution of infection   41
1.3.3 Concluding remarks   48
References   49
Further reading   49

1.4 Contagion and local fragilities in Bergamo and the Seriana Valley

Elisa Consolandi and Marta Rodeschini

1.4.1 Premise   51
1.4.2 The province of Bergamo as a case study   52
1.4.3 The Seriana Valley hotspot   56
1.4.4 Conclusions   60
References   60
Further reading   61

Chapter 2

Mortality and severity of contagion

2.1   Estimation of mortality and severity of the Covid-19 epidemic in Italy

Ilia Negri and Marcella Mazzoleni

2.1.1 Introduction   65
2.1.2 Mortality in Italy in March 2020   68
2.1.3 Mortality estimation for Covid-19   70
2.1.4 Analysis of mortality by age   74
2.1.5 Conclusions   76
References   77

2.2   Mortality and severity of infection in Lombardy

Ilia Negri and Marcella Mazzoleni

2.2.1 Introduction   79
2.2.2 Analysis of mortality data in Lombardy   80
2.2.3 Covid-19 mortality estimate   82
2.2.4 Analysis of mortality by age   85
2.2.5 Conclusions   87
References   87

Chapter 3

Mobility and urbanization

3.1 Commuting in Europe and Italy

Alessandra Ghisalberti

3.1.1 Commuting between proximity and reticularity   91
3.1.2 The European context of mobility   92
3.1.3 A focus on commuting in Italy   95
3.1.4 Conclusions   97
References   97
Further reading   98

3.2 Urbanity and commuting in Lombardy

Emanuela Casti

3.2.1 Urbanity and commuting for reflexive mapping   99
3.2.2 Monitoring commutes in Lombardy   103
3.2.3 The rhizome-like form of commuting in Lombardy   106

3.2.4 Conclusions   109
References   110
Further reading   110

Chapter 4

Pollution and territorial diffusion of contagion

4.1 Correlation between atmospheric pollution and contagion intensity in Italy and Lombardy

Fulvio Adobati and Andrea Azzini

4.1.1 Introduction   113
4.1.2 Links between pollution and contagion   114
4.1.3 Initial assessment   123
References   124

Chapter 5

Dynamics of contagion and fragility of the healthcare and welfare system

5.1 Epidemic onset and population and production density of outbreaks in Lombardy

Andrea Brambilla

5.1.1 Premise   127
5.1.2 Outbreaks and sporting events in Lombardy   128
5.1.3 The production and logistics fabric underlying Lombardy's outbreaks   132
5.1.4 Reticularity between outbreaks in Northern Italy   137
5.1.5 Initial results   140
References   140
Further reading   141

5.2 Nursing and Residential Care Facilities (RSA) and contagion-related fragilities in Italy

Marta Rodeschini

5.2.1 The role of nursing and residential care facilities during the Covid-19 epidemic   143
5.2.2 Elderly care and nursing and residential care facilities (RSAs) in Italy   146

5.2.3 Conclusions   151
References   152
Further reading   153

5.3 The Italian health care system and swab testing

Marta Rodeschini

5.3.1 The healthcare system in Italy   155
5.3.2 The epicenter regions of the epidemic   161
5.3.3 Conclusions   166
References   166
Further reading   167

5.4 Dysfunctions and inadequacies in Health Districts and Nursing and Residential Care Facilities for the elderly in Lombardy as highlighted by the Covid-19 epidemic

Emanuele Garda

5.4.1 A model for elderly residential hospitality to the test   169
5.4.2 The national and Lombard care-system for non-self-sufficient people   172
5.4.3 Lombardy in the plural   175
5.4.4 Conclusions   179
References   181

Chapter 6

Public policies for epidemic containment

6.1 Public policies for epidemic containment in Italy

Fulvio Adobati, Emanuele Comi, and Alessandra Ghisalberti

6.1.1 Premise   185
6.1.2 Contagion containment measures in Europe and their space–time evolution   186
6.1.3 The Italian legal system   190
6.1.4 Italian measures in a space–time perspective   199
6.1.5 Conclusions   201
References   202
Further reading   203

6.2 Containment measures in relation to the trend of infection in Italy

Fulvio Adobati, Emanuele Comi, and Alessandra Ghisalberti

6.2.1 Introduction   205
6.2.2 Main public measures for infection containment in Italy   206
6.2.3 Results of containment measures on infection progress   207
6.2.4 Outcomes of containment measures on citizens fined by the police   209
6.2.5 A difficult balance between Regions and the State: healthcare and transport   210
6.2.6 Conclusions   215
References   215
Further reading   216

Conclusions

Towards spatial vulnerability management for a new "happy" living

Emanuela Casti

1 Territorial fragilities and containment interventions   217
2 Territorial regeneration and potentials set forth by the pandemic   219
3 The anthropocene era: Environmental and pandemic crises   220
4 Happy living   224
References   225

Index   227

List of figures

FIG. 1	Stages of virus propagation at different speeds in relation to social and territorial factors.	5
FIG. 2	Theoretical grounding and spatial dimension of the virus.	8
FIG. 3	Spatial data for analysis of the Covid-19 contagion.	9
FIG. 4	Spatial–temporal spread of the Covid-19 epidemic in Italy.	10
FIG. 1.1	Europe: national distribution of Covid-19 infection from 24 February to 30 June 2020 (absolute data).	20
FIG. 1.2	Europe: distribution of Covid-19 infection as of 30 June 2020.	23
FIG. 1.3	Distribution of urban areas and population density in Europe.	26
FIG. 1.4	Italy: quantification and evolution of Covid-19 infection (absolute data).	31
FIG. 1.5	Evolution of the Covid-19 contagion index in Italy.	33
FIG. 1.6	Lombardy: distribution of the Covid-19 infection in relation to the resident population as of 23 March 2020.	42
FIG. 1.7	Lombardy: evolution of the Covid-19 infection in relation to resident population from 23 March to 14 April 2020.	44
FIG. 1.8	Lombardy: municipal distribution of Covid-19 infection from 24 February to 14 April 2020.	47
FIG. 1.9	Province of Bergamo: evolution of contagion relative to resident population.	54
FIG. 1.10	Province of Bergamo: distribution of contagion from 24 February to 14 April 2020.	56
FIG. 1.11	Seriana Valley: distribution of the Covid-19 infection (absolute data) from 24 February to 14 April 2020.	58
FIG. 1.12	Seriana Valley commutes: daily flows for study and work.	58
FIG. 2.1	Mortality rates in March 2020 for Covid-19 and Covid+ by region (estimates from ISTAT mortality data).	73
FIG. 2.2	The number of deaths and mortality rate for Covid+ in March 2020.	74
FIG. 2.3	Mortality rate for Covid+ by region and by age in March 2020.	75
FIG. 2.4	Covid-19 and Covid+ mortality rates in March and April 2020 by province (estimated from ISTAT mortality data).	83
FIG. 2.5	Number of Covid+ deaths and mortality rate in March and April 2020.	84
FIG. 2.6	Number of deaths and mortality rate for each municipality of Lombardy in March and April 2020.	85
FIG. 3.1	Commuter index on active population members and anamorphic map of residents in Europe.	93
FIG. 3.2	European cities and areas of work-related mobility, with a detailed cut-out view of Milan and Rome.	94
FIG. 3.3	Commuter index on active population members and anamorphic map of municipalities based on resident population in Italy.	96
FIG. 3.4	Index of commuter flows in Lombardy compared to the resident population.	104
FIG. 3.5	Relational intensity of Local Labor Systems in Lombardy relative to active population.	105
FIG. 3.6	Big data and work commutes in Lombardy: a rhizome-like movement.	107
FIG. 3.7	Work Commutes: (A) the nodes of new "communities"; (B) "Community" or "Rhizomes."	108
FIG. 4.1	Europe: distribution of nitrogen dioxide pollution, on average values recorded in January 2020.	115

FIG. 4.2   Europe: different dioxide concentrations before and during restrictions for Covid-19
                pandemic.   116
FIG. 4.3   (A) NO₂: annual legal limit exceedances in urban areas; (B) PM_{10}: monitoring
                stations and daily limit exceedances for health protection.   118
FIG. 4.4   Lombardy: distribution of nitrogen dioxide in the atmosphere in 2019.   119
FIG. 4.5   Lombardy: atmospheric pollutants and allowed daily limit exceedances for
                municipalities.   120
FIG. 4.6   Normalized cumulative indicator of 2019 annual average readings for PM_{10} and
                $PM_{2.5}$, relative to established legal limits.   121
FIG. 4.7   Distribution of RIR and AIA businesses and annual average readings of PM_{10} in the
                province of Bergamo.   123
FIG. 5.1   Reticularity between outbreaks in Northern Italy.   132
FIG. 5.2   Population and companies density in the province of Lodi (Codogno and
                Casalpusterlengo).   133
FIG. 5.3   Population and companies density in the Seriana Valley outbreak (Bergamo).   136
FIG. 5.4   RSA distribution across the Italian regions in relation to elderly inhabitants. 149
FIG. 5.5   Distribution of RSA and old-age indices by region.   150
FIG. 5.6   Evolution of regional number of Covid-19 infected in relation to the swabs carried
                out.   158
FIG. 5.7   Region swabs outcomes and contagion index from March 23 to May 20, 2020.   159
FIG. 5.8   Evolution of the infection by age range in Italy from March 23 to May 20, 2020. 160
FIG. 5.9   Swabs carried out in Lombardy from February 24 to May 4, 2020.   165
FIG. 5.10   Swabs made in Veneto from February 24 to May 4, 2020.   165
FIG. 5.11   Distribution of RSA in Lombardy, scaled to the number of beds.   174
FIG. 6.1   Different containment measures in European Countries between February and
                June 2020.   189
FIG. 6.2   Evolution of the containment measures in Italy in February and March 2020.   200
FIG. 6.3   Number of positive cases on swabs carried out in Italy, between 23 February and
                30 June 2020.   208
FIG. 6.4   Number of fined or reported individuals over 1000 checks in Italy (11 March–30
                June 2020).   209
FIG. 6.5   Establishment of "red zones" and management of transport via regional ordinances
                in spring 2020.   214

List of contributors

Editor

Emanuela Casti is a emeritus professor of geography at the University of Bergamo where she is president of the Centro Studi sul Territorio (CST) and the Diathesis Lab (www.unibg.it/diathesis). Considered an innovator in cartographic theory studies, she has formalized a semiotic theory that investigates the relationship between cartography and geography, extending it to the new systems of cybercartography described in the book *Reflexive Cartography*, published in the Modern Cartography Series by Elsevier. Her researches concern both historical (prehistoric cartography in Valcamonica, renaissance and modern cartography of Venice, and Italian and French colonial cartography in Africa) and contemporary contexts: mapping of both movement and conservation (http://multimap.unibg.it/; https://orobiemap.unibg.it/) for governance (www.bgopenmapping.it; www.bgpublicspace.it; https://cittaaltaplurale.unibg.it/) and regeneration of urban peripheries (www.rifoit.org). She has published more than a hundred essays, which include the books *Reality as Representation: The Semiotics of Cartography and the Generation of Meaning* (Bergamo University Press, 2000) and *Reflexive Cartography. A New Perspective on Mapping* (Elsevier, 2015).

Assistant editors

Fulvio Adobati, associate professor of urban planning, Department of Engineering and Applied Sciences, University of Bergamo

Ilia Negri, associate professor of statistics, Department of Economics, University of Bergamo

Other authors of chapters

Andrea Azzini, cartographer at CST, University of Bergamo

Andrea Brambilla, MA student in geourbanistics and research collaborator at CST, University of Bergamo

Emanuele Comi, post-doc fellow in law, Department of Law, University of Bergamo

Elisa Consolandi, PhD student in geography and research collaborator at CST, University of Bergamo

Emanuele Garda, senior lecturer in urbanistics, Department of Engineering and Applied Sciences, University of Bergamo

Alessandra Ghisalberti, associate professor of economic-political geography, Department of Foreign Languages, Literatures and Cultures, University of Bergamo

Marcella Mazzoleni, post-doc fellow in statistics at the Centro sulle dinamiche economiche, sociali e della cooperazione (CESC), University of Bergamo

Marta Rodeschini, PhD student in geography and research collaborator at CST, University of Bergamo

Acronyms

ADI	Assistenza Domiciliare Integrata—Integrated home care
AIA	Autorizzazione Integrata Ambientale—Integrated environmental authorization
APPA	Agenzia Provinciale di Protezione dell'Ambiente—Provincial environmental protection agency
ARCA	Azienda Regionale Centrale Acquisti—Central purchasing company
ARPA	Agenzia Regionale di Protezione dell'Ambiente—Regional environmental protection agency
ASL	Aziende Sanitarie Locali—Local health authorities
ASST	Aziende Socio Sanitarie Territoriali—Territorial social healthcare companies
ATECO	Attività economiche—Economical activities
ATS	Azienda per la Tutela della Salute—Company for the protection of health
CESC	Centro sulle dinamiche economiche, sociali e della cooperazione—Center on economic, social and cooperation dynamics
CFR	case mortality rate
COM	Communication from the Commission to the Institutions
COVID+	direct and indirect causes of COVID-19
CST	Centro Studi sul Territorio—Territorial Studies Center
DPR	Decreto del Presidente della Repubblica—Presidential Decree
DGR	Delibera Giunta Regionale—Regional Government Decree
DPCM	Decreto del Presidente del Consiglio dei Ministri—Decree of the President of the Council of Ministers
DPI	Dispositivi di Protezione Individuale—Individual protective equipment
DSE	Data Science and Engineering
EC	European Commission
ECDC	European Centre for Disease Prevention and Control
EEA	European Environment Agency
EPFL	Ecole Polytechnique Fédérale de Lausanne
ESA	European Space Agency
ESPN	European Social Policy Network
EU	European Union
EUPOLIS	Istituto superiore per la ricerca, la statistica e la formazione—Regional Institute for Policy Support in Lombardy
FIB	Federazione Italiana Bocce—Italian Bowls Federation
FIGC	Federazione Italiana Giuoco Calcio—Italian Football Federation
GP	General Practicioner
HT	health district
ICF	international classification of functioning, disability and health
ICU	Intensive Care Unit
IFR	death rate from infection
ILI	influenza-like illness
INPS	Istituto Nazionale della Previdenza Sociale—National Social Security Institute
INSEE	Institut National de la Statistique et des Études Économiques

IPPC	Integrated Pollution Prevention and Control
IR	infection rate
ISPRA	Istituto Superiore per la Protezione e la Ricerca Ambientale—Italian Institute for Environmental Protection and Research
ISS	Istituto Superiore di Sanità—Italian Higher Health Institute
ISTAT	Istituto Statistico Nazionale Italiano—Italian National Institute of Statistics
LEA	Livelli Essenziali di Assistenza—Essential levels of assistance
LPS	Laboratorio di Sanità Pubblica—Public health laboratory
LTC	long-term care
MTA	advanced automotive solutions
NASA	National Aeronautics and Space Administration
NHS	National Health Service
NO	nitric oxide
NO_2	nitrogen dioxide
NUTS	Nomenclature of Statistical Territorial Units
OECD	Organization for Economic Cooperation and Development
OxCGRT	Oxford COVID-19 Government Response Tracker
PFR	population death rate
PM_{10}	particulate matter 10 µm or less in diameter
$PM_{2,5}$	particulate matter 2.5 µm or less in diameter
PMI	Piccole Medie Imprese—Small and medium business
PM_X	particulate matter
PPE	Personal Protective Equipment
QGIS	Quantum Geographic Information System
RIR	Rischio di Incidente Rilevante—Companies at risk of major accident
RPE	Respiratory protective equipment
RSA	Residenze Sanitarie Assistenziali—Nursing homes and residential care facilities
SME	Small and Medium Enterprise
SPA	Società Per Azioni—Joint-stock company
SAD	Servizio di Assistenza Domiciliare—Home care service
SARI	severe acute respiratory infections
SARS-CoV-2	severe acute respiratory syndrome coronavirus 2
SME	Small and Medium Enterprise
SIMA	Società Italiana di Medicina Ambientale—Italian Society of Environmental Medicine
SLL	Sistemi Locali del Lavoro—Local labor systems
SSN	Sistema Sanitario Nazionale—National health system
TU	Testo unico—Consolidated law
TEU	Treaty on European Union—Trattato sull'Unione Europea
TUE	Trattato sull'Unione Europea—Treaty on European Union
UEFA	Union of European Football Associations
UFP or UP	ultra fine particulate matter
ULSS	Unità Locali Socio Sanitarie—Local social health units
UN	United Nations
UNESCO	United Nations Educational, Scientific and Cultural Organization
VAT	value added tax
WHO	World Health Organization

Preface

In the 14th century, the Lombardy region of Italy was the first area in Europe to be affected by the Black Death. In 2020, it was again the first region in Europe to be affected by COVID-19. Both pandemics came to Lombardy from China. Within Lombardy, the city of Bergamo was the epicenter of the infection, which then spread to other regions of Italy, Europe, and the rest of the world with continuing deadly effects.

This book presents a detailed case study of COVID-19 by a group of researchers living at the heart of the outbreak in Bergamo. Geography, and location-based analyses and technologies, is central to increasing our understanding of the current pandemic, which, despite the mitigating impact of vaccines, is likely to be with us for some considerable length of time. Professor Casti and her colleagues make an important and unique contribution to the ongoing understanding of COVID-19 and also provide guidance to the actions required to deal with more effective longer-term responses.

Much of the current geographical literature on COVID-19 tends to deal with GIS technologies and how those can outline the importance of location; but as Professor Casti rightly points out, these technologies, although valuable and useful, are not in themselves enough to explain the very complex set of relationships in understanding COVID-19. The book asks the very basic question of "why in Bergamo" and provides a detailed and comprehensive answer to this central question.

The pandemic is treated as a social phenomenon, and to understand it requires a detailed description of the role of territoriality, which the book provides in a comprehensive, reflexive, and interdisciplinary manner. In analyzing and describing the various elements of territoriality and their relationship to COVID-19, this book uses an innovative combined application of maps and textual approach built on the basis of "reflexive cartography." Various aspects of territoriality are considered, including information from a wide range of sources from a time/space perspective. The cartography here is much more than illustrative. It is an integral part of the territorial analysis, which is central to the book and includes elements of the paradigm of cybercartography, such as a multidisciplinary team approach and extensive use of multimedia formats in addition to maps. The editorial team includes a statistician and an urban planner, and the chapter authors include input from a number of younger scholars.

The book concludes with a thoughtful analysis of possible new directions for territorial governance, which argues for an approach that addresses pandemic, environmental, and social issues as an integrated whole.

This book is a valuable addition to the Modern Cartography series, and I congratulate everyone involved in its production.

D.R. Fraser Taylor
Ottawa, February 2021

Editors' comments: The Covid-19 epidemic in Italy, a European epicenter

Emanuela Casti, Fulvio Adobati, and Ilia Negri

This book analyzes contagion trends and the spread of Covid-19 in Italy in relation to territorial features with a focus on the region of Lombardy, which was the most severely affected. Research began in February 2020, as a group of researchers from the University of Bergamo began monitoring in Italy the first epidemic "wave" of viral infection coming from China, which subsequently spread to Europe and the entire world. Monitoring ceased at the end of June, when the epidemic entered an endemic phase and lockdown measures adopted to contain it were finally lifted.

This initial intense period of health emergency in Italy is crucially important for shedding light on the epidemiological dynamics of the disease, for pinpointing potential flaws in our pattern of urban living in times of Covid-19, and, finally, for assessing measures that were put in place in an attempt to stem infection and safeguard the functioning of the health-care system. As also witnessed at present, subsequent Covid-19 waves provide datasets that differ from first-wave data, since infection monitoring has by now extended to increasingly larger population groups, which also include asymptomatic people.[a] Furthermore, later epidemic waves depend on containment and tracing measures that were adopted as well as on citizens' acceptance of restrictions over individual freedoms: all that makes it difficult for researchers to examine contagion dynamics in relation to socioterritorial features.

This study should be placed in the context of social research that developed in the course of 2020 alongside biomedical studies, and it is rooted in the influential claims of one of the famous 19th-century progenitors of German pathology, Rudolf Virchow, who stated, "An epidemic is a social phenomenon that involves some medical aspects."[b] In addition, our research is not so much and not merely an account of the first viral wave that swept across Europe. Rather, it is a territory-focused reflection on a complex issue, an investigation the ultimate goal of which is to derive useful guidelines on how to defend ourselves from subsequent Covid-19 waves or subsequent pandemics.[c]

The initial purpose of this research was to answer the question, "Why in Bergamo?," that is, to investigate why the contagion spread with such unparalleled virulence and gravity in the

---

[a] In fact, in the second wave, the word "outbreak," covering one or more symptomatic subjects testing positive to Covid-19, was replaced by the term "cluster," that is, a group of asymptomatic subjects who test positive during checks aimed at specific groups, such as tourists returning from holidays.

[b] The quote comes from the recent volume by Bernard Henry Lévy, published in English with the title "The Virus in the Age of Madness" (Lévy, 2020).

[c] References to Europe occur in many chapters, but do not amount to a full-scale comparison claim.

province of Bergamo, and later affected much of the region to which this territory belongs, that is, Lombardy.

In order to pursue this goal, besides official infection data made available by the Ministry of Health, we relied on a "toolbox," which included datasets produced over the years on the socioterritorial aspects of the region and used cartographic and geographic equipment from the CST-DiathesisLab to visualize data and translate datasets into information.[d]

However, we soon came to the realization that we needed to formulate a clear starting hypothesis and to lay out a solid theoretical framework on which our analytical research method could be based. Our hypothesis eventually focused on the existence of a relationship between epidemiological features and physical and social aspects of places. Thus, we embraced the notion that *territory affects contagion and that territorial features impinge on the onset, course, intensity, and severity of contagion*. This involved assuming territory not exclusively in its localized dimensions, but rather in relation to its physical and/or social features. In order to address the territorial phenomena of globalization, we needed to adopt a theoretical model, and the reticularity model seemed particularly apposite. A reticular model succinctly states that *in the contemporary, mobile and urbanized world, living unfolds along the intertwining nodes and connections produced by the dynamism of inhabitants both locally and globally* (Lévy, 2008). Unsurprisingly, in times of a pandemic, reticularity of this kind marks ideal conditions for a viral spread. Under such conditions, a contagion will occur both by *proximity*, which results from gatherings or crowding around high-connectivity places, such as *hyperplaces* (public spaces typical of high population density), and by *reticularity*, which derives from people's movements on collective public transport (Lussault, 2007, 2017).

Buttressed by recent scientific developments on the role played by spatial dimension in any social phenomenon (in the wake of a *spatial turn*) and by the novel approach to mapping as a *medium* for spatial representation, our research pursued a twofold objective, namely, (i) to analyze the contagion in its space–time dimension, highlighting territorial vulnerabilities and envisaging actions for overcoming them under the inevitable aegis of ecological awareness and (ii) to build reflexive maps, which, against an alarmist plan of massive reliance on GIS maps on the epidemic issued at various levels (institutional or otherwise), would instead favor a vision capable of consciously facing challenges the contagion poses.

The resulting analytical method was a real-time monitoring of infection across various territories analyzed at multiple scales (municipal, provincial, regional, national) with a view to pinpointing epidemic differences. Such data were presented using a cartographic model that cross-referenced them to socioterritorial data with the aim to facilitate interpretation and process new information, subsequently laid out in written form. In fact, cross-referencing of territorial aspects (population distribution and composition; citizen mobility; distribution of production facilities and pollution; and reaction capacity of the health-care and welfare

---

[d]Over the years, the Diathesis Cartographic Lab, within the Territorial Studies Center at the University of Bergamo, has experimented with new forms of *digital mapping* for diagnostic research at local, national, and international levels. These were aimed at not only territory and landscape designs but also at urban and environmental governance. They applied innovative theoretical models (Casti, 2000, 2015), methods that tapped data from multiple sources (for instance, land surveys and observations, statistical and archival analyses, as well as social media and big data), and digital platforms for data collection, processing, interpretation, and disclosure (www.unibg.it/diathesis).

systems) and epidemic features has made information more complete and more reflexive, as envisioned by the paradigm of *cybercartography* (Fraser Taylor et al., 2019).

The coupling of maps and textual descriptions is also a distinctive feature of this book, which combines written and drawn text to explicitly put forth a model of "unlimited semiosis," an additive text–map relationship. This model complies with the conventional ranking of text over map, without however excluding the possibility of a mutual implication between the two: text and figure not only refer to each other but also mutually support each other, thus legitimizing the overall relevance of the information they convey via such a dense network of cross-references.[e]

Data obtained through field surveys, multiple information sources, and interdisciplinary meetings aimed at broadening our view of the phenomenon we were experiencing were systematized. They showed that in its first wave, the Covid-19 contagion in Italy struck and spread differently in relation to the physical and social factors of the territory. The epidemic epicenter was located in the north, in the Po Valley, where it remained throughout the entire spring wave of Covid-19. On the contrary, there was, a tripartite zoning of the country into what we named the *Three Italies*: the first included the greatest part of northern provinces, where infection intensity and gravity were the highest; the second included most of the remaining provinces of the Po Valley and some neighboring areas, where contagion intensity ranged from medium to high; and the third comprised the rest of the nation, which was only mildly affected.[f] This consolidated our research approach and confirmed what also seems to emerge from data on international contexts, namely, epidemic features delineate an *anisotropic space that differs locally, regionally, and nationally*.

The fact that Italy, like the rest of Europe, is currently experiencing a second pandemic "wave" does not diminish but rather increases the relevance of these observations. They can be applied comparatively, and thus encourage a reflection on how we may act to consciously address the challenges posed by what some authors have named our entry into the Anthropocene. Outlined as a new geological era, the Anthropocene is characterized by the environmental crisis currently affecting earth. In that context, the SARS-CoV-2 virus pandemic marks the first global increase in awareness of the need to seek a new model of life and living (Lussault, 2020). In times of the pandemic, there emerged, on the one hand, the need to rethink the spatial practices of our urban and mobile living (i.e., the spatiality of social phenomena), the relationship between individuals (the interspatiality of inhabitants), and the modes of cohabitation (the way in which societies spatially organize coexistence between individuals). On the other hand, social policy has shown that even though they possess the knowledge and technological and organizational skills to stem contagion while

---

[e]Other forms of communication, such as video clips and media press releases in multiple formats (such as social networks, online channels, television, newspapers, etc.), bimonthly reports, and journal articles in Italy and abroad were also added. The underlying aim was to convey research outcomes in real time.

[f]In the second wave, on the other hand, contagion has also extended to most of the Peninsula: socioterritorial factors behaved differently than in the first wave and must be ascribed mainly to a generalized inadequacy of the health-care and welfare systems. A comparison between infections in April 2020, when the Covid-19 epidemic peaked, and in October 2020 shows that all Italian regions, even those in the south and the islands, are affected (albeit to a lesser extent), with the exception of the large metropolitan areas of Rome, Naples, and partly Palermo, whose infection factors differ from the ones outlined for the Po Valley.

guaranteeing basic needs, complex societies such as western democracies have in fact exacerbated social disparities that are typical of a capitalist model.[g]

In light of this, our book sets out to propose a generalizable reflection on how to face contemporary challenges without neglecting ethical and social challenges.

At this point, we should mention where and when the study was carried out. We clarified above that it all started as a field analysis carried out in Bergamo, the epidemic "eye of the storm," at the beginning of the pandemic. Geography was summoned from the start as a connecting discipline among the many others called upon to act in the resolution of complex social phenomena.

As evidence of this, it should be remembered that this study was initially conceived and coordinated by the geographer Emanuela Casti, at the Territorial Studies Center, and was later picked up by other cocoordinators, such as Fulvio Adobati, an urban planner, and Ilia Negri, a statistician, both tied to the University of Bergamo. While preserving the core of research, that is, a geographical investigation of Covid-19 infection, Adobati and Negri broadened its interdisciplinary perspective with the aim to better detail the relevance attributed to the spatial dimension of social phenomena. Thus, interdisciplinarity is a core value attested by the results achieved and the intertwining of aspects under investigation.

Another qualifying feature is the presence of two generations of researchers in the research team: the first set up the group, thanks to experience in the field, critical acumen, and theoretical reflection, and the second, or "digital generation," contributed a more competent mastery of IT tools, and, thanks to this unforeseen experience, learned that a knowledge of means does not suffice in order to reach a goal.[h]

The volume cohesively unfolds in chapters written by different authors around one major research proposition, namely, the intensity and severity of epidemic spread are influenced by territory and precisely by the presence/absence of some geophysical and/or social aspects. We iterate these once more: population density and composition, mobility and commuting, air pollution, and health-care and welfare systems.

Theoretical grounding is discussed in the Introduction, and issues of theory are subsequently picked up in turn by each chapter, in focused analyses of the spatiotemporal evolution of a multiscale epidemic in Italy, visualized via reflexive mapping. In detail, the Introduction lays out the theoretical and methodological grounds of research and the role taken up by cartography in communicating contagion aspects in an interdisciplinary context. The new reflexivity paradigm in *digital mapping* and the communicative implications of digital constructs are outlined in order to pinpoint, in our specific case, territorial fragilities on which to act to counter the epidemic crisis.

The first chapter deals with the distribution of the infection in relation to population. It comprises four sections and initially proposes a Europe-wide overview of infection by

---

[g]Street riots that are taking place in various parts of the western world bear witness to social unease: we need to consider that in times of crisis those who are the most disadvantaged are precisely the weakest sectors of the population.

[h]Since the beginning, the latter have taken part in research on a voluntary basis, driven by a concern for ethics and convinced that research was not detached from the events of life, but rather was a defense tool for understanding and improving it. Their names are Fulvio Adobati, Andrea Azzini, Andrea Brambilla, Elisa Consolandi, Emanuele Garda, and Marta Rodeschini.

cross-referencing epidemic data with those on population distribution and composition of the population. Subsequently, it examines the evolution of the contagion in Italy, highlighting how Lombardy was the only region to be severely hit. The third section deals with epidemic analysis in Lombardy from both a quantitative and an evolutionary point of view, identifying the "backbone" of maximum contagion in the eastern part of the region, which connects the two epidemic outbreaks of Lodi and Bergamo. Finally, the fourth section zeroes in on the local level, recalling physical and socioterritorial factors in the Seriana Valley, which made headlines worldwide for its sad record as the most devastating outbreak in Italy.

From a purely statistical perspective, but always with special attention to spatial features, the second chapter investigates mortality and contagion severity in Italy and in Lombardy. It highlights the inadequacy of publicly issued data on the outcomes of infection and demonstrates that during the most serious and critical phase, official deaths due to Covid-19 were in fact lower than the actual number of people who died of the disease. It then sets forth an estimation method based on excess mortality observed in the area. What emerges is that deaths due to an implicit Covid-19 cause or Covid-related illness hit Italy harder than other European nations, struck the region of Lombardy more severely than other Italian regions, and that, among the provinces, Bergamo is the one that paid the most in terms of mortality. That confirms the hypothesis of an anisotropic territory outlined by the disease and of a subdivision into three macroareas both nationally and regionally.

The third chapter outlines the European scenario in relation to major mobility corridors and addresses the Italian situation. It identifies Lombardy as the region with rhizome-like commute patterns, which affect the spread of infection by facilitating reticular contact and interaction between people. The use of Big Data related to mobility flows makes it possible to process reflexive *mapping*, which conveys their reticularity and their spatial and temporal concentration, a condition which proved to be particularly dangerous with regard to collective public transport.

The fourth chapter analyzes atmospheric pollution, carrying out a detailed mapping at the regional level for Lombardy, also in relation to the Italian national context and the European context. The analysis is based on the assumption that there exists a relationship between the degree of salubrity of settled environments and possible vulnerabilities in the health of inhabitants. Specifically, the diffusion of two major atmospheric pollutants, nitrogen dioxide and atmospheric particulate matter (PM_{10} and $PM_{2.5}$), is analyzed. The purpose of this section is to provide assessment criteria by collecting clues (with the aid of relevant cartographic representations), which may serve to verify potential correlations between levels of environmental pollution and the intensity and virulence of territorial contagion.

The fifth chapter explores the dynamics of viral diffusion and addresses territorial factors that facilitated contagion, starting with outbreaks initially recorded in the Po Valley and moving on to the later epidemic phase. Diffusion dynamics (both by reticularity and by proximity) are evaluated on sporting events data that involved communities in the period prior to the discovery of outbreaks. The chapter also examines the health-care and welfare systems in Italy, looking for factors in their setup or management that may have affected the intensity and severity of the viral spread. Finally, the analysis zeroes in on Lombardy to show that both the health-care and the welfare systems rely on centralized facilities that have made it difficult to control the virus.

The sixth chapter provides a review of contagion-containment measures adopted by EU institutions and by various European states. It then develops an analysis of the Italian national context, also as played out at the regional level, with comparative references to provisions adopted in other European nations. Thus, elements are offered to identify possible correlations between infection curves and public measures adopted on a national and regional scale and to sketch a cautious assessment of effectiveness of such measures.

The concluding section completes the circle by summarizing the distinctive features of the research projects and the results achieved. The current health crisis has led to an acknowledgment of social and territorial vulnerabilities: these bring into question current models of living and urge us to rethink them on the basis of the vulnerabilities this research has exposed. Pandemic, environmental, and social issues are but three inseparable clusters we need to address in order to rework territorial governance. We need a new model based on the recognition of a much-needed overhaul of environmental quality profiles and services. We also need to envisage a progressive transition toward renewed ways of living territory, which may be seen as instruments for fulfilling the needs of humans, ensuring their quality of life and, ultimately, achieving well-being and happiness.

References

Casti, E., 2000. Reality as Representation. The Semiotics of Cartography and the Generation of Meaning. Bergamo University Press, Sestante, Bergamo.

Casti, E., 2015. Reflexive Cartography. A New Perspective on Mapping. Elsevier, Amsterdam.

Fraser Taylor, D.R., Anonby, E., Murasugi, K. (Eds.), 2019. Further Developments in the Theory and Practice of Cybercartography. Elsevier, Amsterdam.

Lévy, J., 2008. Un évènement géographique. In: Lévy, J. (Ed.), L'invention du monde. Une géographie de la mondialisation. Presses de Sciences Po, Paris, pp. 11–16.

Lévy, B.H., 2020. The Virus in the Age of Madness. Yale University Press, New Heaven-London.

Lussault, M., 2007. L'Homme spatial. La construction sociale de l'espace humain. Seuil, Paris.

Lussault, M., 2017. Hyper-lieux. Les nouvelles géographies de la mondialisation. Seuil, Paris.

Lussault, M., 2020. Chroniques de géo' virale. Ecole urbaine de Lyon/Editions deux-cent-cinq, Lyon.

Introduction: Territorial analysis and reflexive mapping on the Covid-19 infection

Emanuela Casti

1 Context

On February 24, 2020, when the Italian media announced Italy's entry into an epidemic phase from Covid-19, territorial analysts believed the matter did not concern them: The epidemic was a biomedical issue and researchers only involved doctors, virologists, or epidemiologists. However, evident territorial differences that the virus presented in its expansion and spread soon outlined a differentiated and anisotropic epidemic space, which attracted their attention. After a few weeks, when the region of Lombardy became the European epicenter of Covid-19, doubts vanished: Territorial analysts were to enter the field and provide their *expertise* to try to understand what was happening in their territory.

Faced with a new and unprecedented challenge, a group of researchers in Bergamo set up a study based on three features, namely, implementing a database on socio-territorial information and on the epidemic, with a view to mapping such information and obtaining spatialized data on which a set of initial hypotheses could be based; taking into account the spatiotemporal characteristics of the Covid-19 spread and interpreting them in light of geographical theories; and applying semiotic theories to the construction of a *reflexive mapping* for facilitating an interpretation of the phenomenon.

This introduction provides a brief account of how research progressed and traces our gradual realization that the epidemic was well beyond the purview of virology: Rather, it was a social event whose spatial implications expose the fragility and the most vulnerable sides of contemporary living.

2 Epidemiological data and the issue of sources

The first research obstacle was the reliability of data on the Covid-19 infection issued by the Italian Ministry of Health after being collected by regions and other territorial bodies (such as provinces, municipalities, and regional health services).[a] As they were issued by territorial

[a]Contagion data used in this research come from the Presidency of the Council of Ministers-Department of Civil Protection (http://opendatadpc.maps.arcgis.com/apps/opsdashboard/index.html#/b0c68bce2cce478 eaac82fe38d4138b1). However, some of these are provided by the Higher Health Institute (*Istituto Superiore di Sanità*) and were made *machine readable* thanks to OnData processing (see https://www.epicentro.iss.it/coronavirus/sars-cov-2-surveillance-data).

agencies whose main responsibilities did not include statistics, such data lent themselves to oversights and relied on a range of different survey methods. Although, as laid out by the Italian Constitution for states of emergency, antiepidemic measures were centralized and issued directly by the Council of Ministers, their applications were managed locally by the regions, with notable differences over testing and preventive or operational measures for disease containment. Processed at the local level by local health authorities (ASLs), data were sent to each region's central department and from there to the Ministry of Health, which was in charge of disseminating them in an aggregate form at the national level. However, as the epidemic raged, difficulties in data transmission arose, that is, information was delayed, dispersed, or was simply incomplete. Moreover, it was interpreted in ways that differed from the diagnostics used. Furthermore, such data applied only to a particular and limited portion of the population, for they referred exclusively to people who were being hospitalized and thus to sick individuals who presented symptoms. Therefore, data were collected upon hospitalization, and the recorded data showed the evolution and outcomes of the disease (such as mortality, severity of intensive care, or otherwise.) Then, these data were assembled with diagnostic information on individuals confined to their homes with mild symptoms or on asymptomatic individuals. In the first months of the epidemic, data assessment occurred only at the hospital level. This meant that even plain cumulative data on daily infections were unreliable, since diagnostics were carried out only on a section of the population and varied in size depending on the swab testing that individual regions performed and on the results that they actually forwarded to the Ministry. This led some researchers to question the reliability of data: to cast doubt on the actual extent of infection and to challenge the legitimate use of such data for deriving exact infection indicators.

However, the objective pursued in our research, which is to investigate contagion in relation to territorial features and not as a virological phenomenon in its own right, places less importance on numerical data accuracy. Our aim was rather to sift such data for clues to outline trends that could be significant even despite epidemiological approximation.[b]

Data dearth was partly compensated by a number of key secondary sources, such as scientific blogs, *collective databases*, shared online platforms, and qualitative data made public by large web operators. Initially, these sources made up for the lack of reports and studies on the epidemic, which various institutional bodies were still assembling. This made it possible to kick-start comparison and complementary information research.[c] A noteworthy instance among the many in Europe is *Medium*, an online platform that enables users to publish

---

[b]This goal was set after due consideration of opinions by biomedical experts, who first interpreted data, and by statisticians, who compensated for the lack or scarcity of data by building models derived from comparative data collected before the time of the epidemic.

[c]Wikipedia or Google have also made available a series of reports on changes in people's travel habits during the pandemic. See https://www.google.com/covid19/mobility/; in addition to institutional data by the Ministry of Health and the Higher Health Institute (2020), http://www.salute.gov.it/imgs/C_17_notizie_4766_0_file.pdf; the Office for National Statistics (ISTAT) and the Higher Health Institute (2020), https://www.epicentro.iss.it/coronavirus/pdf/Rapporto_Istat_ISS.pdf; the Higher Health Institute (2020), https://www.epicentro.iss.it/coronavirus/pdf/sars-cov-2- survey-rsa-report-3.pdf; and the National Environmental Protection Agency (ISPRA) (2018), https://www.isprambiente.gov.it/it/pubblicazioni/stato-dellambiente/xiv-rapporto-qualita-dell2019ambiente-urbano-Edizione-2018.

content also aimed at interdisciplinary scientific dissemination.[d] Another example of communication along these lines is the Groupe d'Etudes Géopolitiques (GEG) founded by the École Normale Supérieure in Paris to advance interdisciplinary reflection on European geopolitics. GEG studies tap data from a range of disciplines (such as geography, philosophy, economics, sociology, and literature) in a transnational and multilingual perspective. They involve several universities and écoles, in particular, the École Polytechnique, Science Po and Paris IV, and the Sorbonne.[e]

Our research also tapped a new technology, such as algorithm-driven processing of Big Data, which proved effective, even though it was mostly applied to a limited number of socio-territorial features, such as mobility or pollution.

Ultimately, the pandemic caught us unprepared with regard to sources of territory-based research and forced us to operate on an analytical basis, which has yet to gain firm grounding. While aware of the possible risks that such a procedure entails, we have deemed it prudent to include it in our analysis of the broader aspects of the contagion and of more circumstantial socio-territorial features. With regard to the former, we believed it was crucial to first address the space–time dimension.

3 Stages of virus propagation in relation to social and territorial factors

As we dwell on this feature of the Covid-19 infection, we will attempt to find investigative leads on the example of what happened in Italy. We should first underline the rapid onset of infection: Discovery of the first Covid patient at a specific location was followed within barely a few hours by reports of other Covid patients elsewhere. At first glance, the reported sites presented no evident similarities. However, they did seem to share at least one localization feature: They were part of peri-urban areas across the great Po Valley conurbation of Northern Italy.[f] This is the most densely populated area in the country, the most economically vibrant, and the one boasting an advanced health-care, diagnostics, and hospitalization system. The territory is divided into regions (namely, Lombardy, Veneto, Emilia-Romagna, Piedmont, Val d'Aosta, and Liguria) administered by different political parties. These are complex features, which we will systematically address in this volume.

If we temporarily disregard the localization of outbreaks and instead focus on the *spatial diffusion* of infection, we are struck by the *speed* whereby the contagion spread.

---

[d]The platform, created in 2017, played an important role during the pandemic by bringing together European researchers and universities. See the *Medium* website, https://medium.com/. Institutions that use this platform include the Ecole Urbaine de Lyon (https://medium.com/@ecoleurbaindelyon), the University of Oxford (https://medium.com/@Oxford_University), and the MIT Technology Review (https://medium.com/mit-technology-review).

[e]In view of the current health emergency, the Medium Group set up *the Observatoire Géopolitique du Covid-19*, which publishes daily data updates on the evolution of the epidemic along with studies, maps, interviews, and reports related to the pandemic. See https://legrandcontinent.eu/fr/observatoire-coronavirus/.

[f]Data related to the insurgence of Covid-19 hotspots in Lombardy suggest a connection, if not an actual reticularity model, between outbreaks, as also attested by the sporting events that took place in the weeks before infection hotbeds were recorded. See Chapter 5 of this volume.

Unsurprisingly, therefore, infection speed will be a crucial factor in our analysis of the pandemic.[g] Infection speed rate is a highly relevant variable for effective control of the epidemic, since, as epidemiologists remind us, Covid-19 is a respiratory infection that occurs by contact between individuals in urbanized and high-mobility environments, where viral propagation is notoriously swift. Therefore, the temporal aspect of the contagion was monitored within our research model by factoring speed in the spatial diffusion of infection in relation to measures being taken to contain it. Following WHO guidelines for identifying epidemic phases,[h] we broke down the first phase of Covid-19 in Italy into three phases, namely, first, an *onset phase*, when first cases were reported and the existence of *infection hotbeds or hotspots* (i.e., tangible threats of epidemic spread to an entire community) was ascertained[i]; second, an *epidemic phase*, when infection peaks were reached; and finally, an *endemic phase*, when the number of infected individuals decreased without disappearing (Fig. 1).

The duration of these phases and the speed of propagation differed as follows:

- The *onset* phase and the localization of *hotbeds* may be placed between February 24, 2020, when the first outbreak was ascertained, and the beginning of March, when the epidemic entered the epidemic phase[j]. It should be stressed that since this first phase, contagion patterns outlined a subdivision of Italy into three distinct zones, with an epidemic epicenter located in the north, in the Po Valley, where it remained throughout the entire spring wave of Covid-19. Subsequently, the contagion spread to neighboring territories in the north, but the intensity and severity of infection recorded in this region did not extend to the rest of Italy[k]. Such a spatial discrepancy led us to investigate a number of climatic and morphological features of the most affected area, which, in addition to setting it apart

---

[g]It is precisely this ability to spread quickly across a large number of territories that led researchers to define the SARS-CoV-2 virus and the resulting Covid-19 infection in terms of *pandemic*. See Jackson (2020).

[h]The World Health Organization (WHO) and the Center for Disease Control and Prevention of the United States (CDC) have codified epidemic phases in these terms: (1) emergence of a new microorganism dangerous to humans; (2) identification of contagion cases; (3) onset of a globally widespread infection; (4) acceleration of the pandemic wave, with an epidemiological curve pointing upward; (5) progressive and steady decrease of cases; and (6) end of the pandemic and at the same time beginning of the preparedness phase to further waves. See https://www.who.int/influenza/preparedness/en/ and https://apps.who.int/iris/bitstream/handle/10665/259893/WHO-WHE-IHM-GIP-2017.1-eng.pdf?sequence=1.

[i]From a territorial point of view, a hotbed should be understood less in terms of the first recorded case of infection than as the site where a number of symptomatic individuals who may be traced back to one suspect case are found: These might spread the virus and threaten an entire community.

[j]Virologists have estimated a duration of between 1 and 2 weeks for this coronavirus phase. In Italy, the phase occurred in sequential periods affecting Northern Italy between the end of February and the beginning of March and, subsequently, the entire peninsula, apart from a few internal regions and the islands, where the contagion emerged a few days later.

[k]Only during the second wave, that is, the fall phase, did contagion spread relatively homogeneously throughout Italy, even though Lombardy held a sad record in that respect for most of the time. See the report published by the Office for National Statistics (ISTAT) and the Higher Health Institute (*Istituto Superiore di Sanità*) in December 2020, available at https://www.iss.it/documents/20126/0/Rapp_Istat_Iss_FINALE +2020_rev.pdf/b4c40cbb-9506-c3f6-5b69-0ccb5f015172? T = 1609328171264.

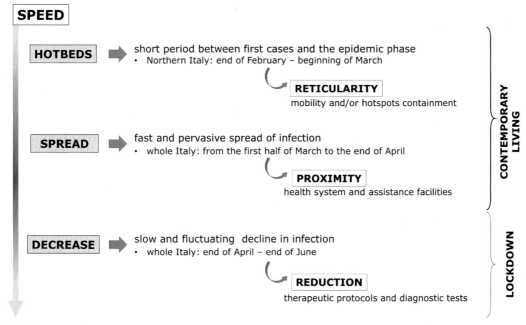

FIG. 1   Stages of virus propagation at different speeds in relation to social and territorial factors.

from the rest of the peninsula, also make it the most polluted region in Europe[l]. Settlement issues would also have to be addressed, given that the Po Valley has a high population density and marked territorial connectivity, which are unique features in the national setup. High people mobility, especially the mobility of commuters whose reliance on collective public transport creates crowding and who remain at close contact indoors even for protracted periods, was an easy clue for explaining viral transmission.

• The fast and pervasive *spread* of the epidemic quickly revealed the severity of Covid-19 and its ability to undermine both health care and normal social and economic functions in Italy. A series of factors affected its progress, including the lack of social distancing among commuters, the difficulty hospitals faced in coping with a high number of patients, the inadequacy of the local health system – most notably in the region of Lombardy – and, finally, the inefficiency of the health-care system for the elderly, which caused a serious and forceful spread of infection in nursing homes and residential care facilities[m]. In Italy, the epidemic phase began in the first half of March, whereas the peak of infection was recorded at the beginning of April and remained high till the end of the month. Even at this stage, the severity of infection recorded in Lombardy, especially in the province of Bergamo, was unmatched elsewhere in Italy.

[l]On the issues of pollution and globalized living in the Po Valley, see contribution in the Chapter 4 of this volume.

[m]Hospital system diversification across Italian regions and in nursing homes (such as the Nursing and Residential Care Facility—RSAs) is discussed at length in Chapter 5 of this volume.

- We finally get to the *endemic* or *decrease* phase, occurring at gradual speeds, and arguably traceable to social-distancing policies and other competing factors. A nationwide lockdown in Italy was decreed on March 9, 2020; its effects led to a slow and fluctuating decline in infection, which lasted until the end of June 2020[n]. At the time, containment strategies were adopted with a view to a progressive reopening by May 18 and without waiting for a complete disappearance of the virus, which virologists believed might be achieved solely by administering vaccines (Jackson 2020). In the meantime, however, the development of therapeutic protocols, the start of the summer season, and the organization and implementation of diagnostic tests favored a reduction in viral circulation and marked the end of emergency.

Overall, the first epidemic wave in Italy covered a period of 4 months (from February 24 to June 24, 2020) with markedly diversified outcomes across the Italian regions attributable to socio-territorial factors. We will dwell on such factors, after the range of empirical data discussed so far has been duly corroborated against a theoretical framework.

4 Theoretical grounding and spatial dimension of the virus

Scientists by now largely agree that in order to understand and interpret social phenomena, one must account for their spatialization: *Where it is* things happen is crucial to understanding *how* and *why* they happen. Although this *spatial turn* has undeniable interdisciplinary repercussions (Warf and Arias, 2009) for coming to terms with the complexity of social reality, for geographers it marks a unique opportunity for reassessing methodological priorities. The Covid-19 epidemic occurred at a time when this perspective had already been verified, and equipped with analytical methods to meet the challenges in the field of Geography.[o] There, the *spatial turn* theory had not been considered exclusively as yet another recognition of the importance of localization in social phenomena, but rather as an acknowledgment of the relevance of geographical analysis itself, which has traditionally aimed to unveil territorial action as the foundation of *territoriality*, that is, of the ways in which a given society functions. In short, the configuration of territories was no longer assumed as a background canvas, but rather as one of the distinctive causes of the contagion, given that the contagion did vary in terms of propagation rate and intensity of the Covid-19 virus across different areas.[p]

Having embraced territoriality as an analysis plan, we also took up the assumptions that *living in the contemporary world is mobile and urbanized* and that it unfolds *in the intertwining nodes and connections* outlined by the dynamic movement of inhabitants, as they endeavor

---

[n]On this issue, see Chapter 6 of this volume.

[o]The responsible pathogen, named SARS-CoV-2, was first reported in the Wuhan area of China in December 2019. It turned into a pandemic in <3 months. It led to *lockdown measures*, which affected half the world's population, in attempts to limit viral spread and to ease pressure on health systems.

[p]Marco Maggioli defines *spatiality* as a set of conditions and practices linked to the placement of individuals and groups relative to each other. Conversely, he defines *territoriality* as a process built on the forms, structures, and assets of a territory, which individuals themselves have helped shape (Maggioli, 2015, pp. 51–66).

to organize their daily lives not only solely in places where they live but also by interaction with global network systems (Lévy, 2008). Therefore, mobility is of great importance in the contemporary world. The dynamics triggered by a continuous flow of people and information amplify and accelerate the space–time cadence between days and nights, weeks and weekends, and the passing of seasons. In turn, altered cadences transform places where individuals live or connect, either in person or virtually. This pattern of living, defined as *urbanity*, is a conspicuous feature of contemporary living. As such, it entails a radical change in the analytical perspective because it shifts our attention from the materiality of territories to the action of their inhabitants (Djaiz, 2020).

It is worth noting that the notion of *urbanity* is not directly or merely related to that of *urban*. Rather, it defines a specific mode of living, not sedentary but mobile, expressed by inhabitants seen as *city users* who are simultaneously connected on multiple scales (Lussault, 2007, 2017). Along similar lines, cities should no longer be viewed in terms of a binary duality of center vs periphery, but rather as systems grounded on mobility, that is, as nodes of a mesh where the dynamics of local and global living intertwine (Hall and Pain, 2006; Soja, 2000). Therefore, the global space of urbanity appears as an extremely pervasive phenomenon, capable of materially integrating within one territory a wide range of systems, such as residential, productive, cultural, and service-related, which differ in terms of density and diversity. Ultimately, people's new mode of inhabiting places takes shape in widespread *urbanization*. This certainly affects contagion, which is amplified by mobility and inevitable crowding in public spaces and becomes the visible expression of social dynamics, whereby a range of interests, services, and modes of experiencing urbanity (Lévy, 2020; Lussault, 2020) converge on a place.

In accordance with this approach, the territorial features involved in the contagion were not considered here individually, but in relation to each other. Hence, population density was analyzed in relation to mobility, pollution was related to morphological and climatic conditions, and health-care systems were viewed in connection with population, hospital types, and care management models. As we addressed such conditions, we also zeroed in on factors that may have favored the contagion, such as persistent pollution and impaired air quality, or on situations that involved crowding, which made isolation difficult: These features were defined as instances of "territorial fragility," which research must adequately address in order to design new habitation models equipped to face epidemics.[q] On the basis of this complex tangle of issues, we set out to interpret the devastating outcomes of the contagion in Lombardy.

From the start, it was clear that all the aspects mentioned above are in fact significantly present in Lombardy. Extensive population density in the metropolitan area of Milan spreads to the rest of the region across many medium or small towns. What ensues is a multidirectional and rhizome-like pattern of commutes, which creates ideal conditions for an intensification of infection.[r] In short, urbanity enables us to take the dynamism of inhabitants and the complexity of urbanization as two sides of the same coin: a territorial system based on

[q]See the Conclusion chapter of this volume.

[r]To be sure, a thorough explanation of Lombardy's very high mortality rate will have to await results of biomedical and epidemiological research still under way. As of now, however, it can be argued that a contributing cause for this outcome was the inadequacy of the health-care system in coping with the very high numbers of infected people and the management of some Nursing and Residential Care Facilities (RSAs).

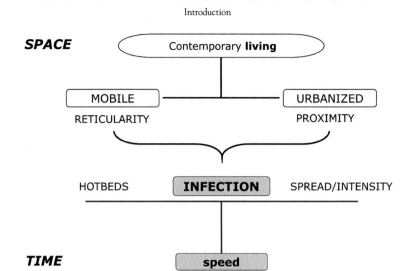

FIG. 2   Theoretical grounding and spatial dimension of the virus.

mobility that amplifies the contagion in times of a pandemic because it promotes crowding in public spaces.

Taken separately, inhabitant mobility or population density may not be classified as factors facilitating the insurgence or the spread of the virus, as we were hastily led to suppose. They do become such, however, if taken in conjunction, when they in fact expose multiple systemic fragilities. Similarly, pollution not only depends exclusively on atmospheric emissions but also involves factors that may prevent pollutant dispersion, such as climatic and/or morphological features. Finally, with regard to the ability to tackle the epidemic and limit its most devastating outcomes, it would be misleading to label and blame health-care or assistance facilities as quantitatively deficient. In fact, it falls seriously short of considering the diverse range of health-care policies adopted by the regions, which in times of an epidemic did make a difference.

In short, adjusting our theoretical scheme to epidemic data for Lombardy enabled us to address the space–time dimension of Covid-19 spread as a key factor, and showed us how to interpret contagion data relative to Italy as a whole (Fig. 2).

5 Geographical implications of the contagion and Italian outcomes

Our analysis was developed along three main axioms, derived from the temporal and dimensional features of the contagion, namely, (i) outbreaks, (ii) diffusion, and (iii) intensity. As research continued, a fourth axiom was added, which was related to the severe manifestation of epidemic contagion in Lombardy (Fig. 3).

The typology and development of a contagion, in relation to both territories and peculiar features of infection, have outlined a space–time map. Such a map highlights the insurgence of the disease in the north and its persistence in the region for the entire duration of the epidemic wave, in ways that are substantially different from the rest of the country.

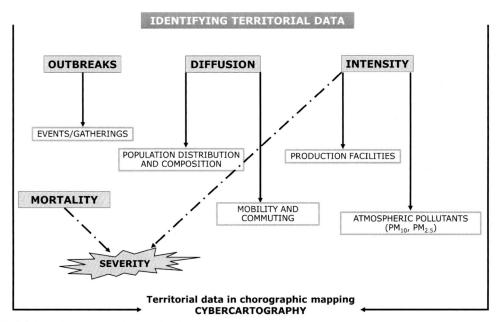

FIG. 3 Spatial data for analysis of the Covid-19 contagion.

The initial outbreaks in Italy were of course identified in the regions of Veneto and Lombardy. However, while the Vo' Euganeo outbreak in the Padua area was easily contained,[s] the Lombard outbreaks of Codogno in the Lodi area and those of Alzano Lombardo-Nembro in the Bergamo area have urged us to reflect on the need for swift action, unhampered by other questions or financial concerns over the cost of localized closures, especially where a given geographical setup makes it possible to contain only the infected area.

Two diffusion dynamics were detected, namely, (1) by proximity, when the virus spreads in neighboring places and (2) by reticularity, where propagation occurs via people who live in different places and come into contact as they move.[t] The Po plain outbreaks, monitored via sporting events, ultimately suggested that both dynamics were in fact present during the first epidemic wave and that they involved small towns more than large agglomerates.

Onset phase data were later cross-referenced to data on the next phase, or the epidemic phase. With an eye on the intensity and gravity of Lombardy's case, we associated the contagion to population composition at the Italian level, thereby obtaining data on the most affected age groups. Of these, the most relevant was a division of the Italian nation into "Three

[s] A small municipality of about 3000 residents based on cluster typology settlements (hamlets) and a secondary road network. The town was restricted and quarantined. Tests were periodically carried out to ascertain infection and study contagion trends, which led researchers to discover a possible viral transmission even via asymptomatic subjects.

[t] In an effort to understand viral spread, researches cataloged events (initially sporting events) that caused gatherings in the weeks immediately preceding the health emergency. See Chapter 5 in this volume.

Italies" or zones, based on the intensity and severity of the contagion reported at the national level. This difference remained unchanged throughout the first epidemic wave, as shown in Fig. 4 and as discussed in detail in the relevant chapter. Italy's three zones outline a geographical model that may not be explained virologically; rather, it depends on spatial differences connected to features such as pollution and rhizome-like commute patterns and to different degrees of urbanity, which call for further study.[u]

In short, the distinctive socio-territorial factors of the Po Valley, not to be found in their complex configuration elsewhere in Italy, have corroborated our initial research hypothesis, namely that, territories influence speed of infection and viral intensity. To be sure, this conclusion does not come directly from knowledge of the virus and is derived from spatial factors of the affected regions. It may not solve the underlying questions or compensate for the lack of virological expertise. However, it does set out a research path that is hard to escape; this path leads us to postulate the incompatibility of contemporary living with pandemics and forces us to look for viable solutions.

At that point, all that remained for our research was to address the need to disclose and disseminate results and to publish the bulk of cartographic documents we had produced in order to reach them. Spatialization of epidemic data and cross-referencing with various socio-territorial features had been achieved via *chorographic mapping:* a set of representations

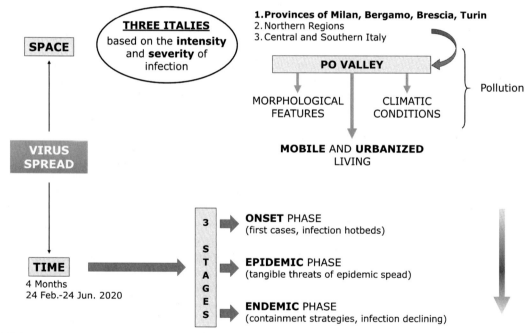

FIG. 4   Spatial–temporal spread of the Covid-19 epidemic in Italy.

[u]This is the first chapter of this volume. An in-depth research is currently being completed in Italy, which involves most of regional cartographic centers in an investigation of these aspects and will be soon printed by the A.Ge.I (Association of Italian Geographers).

that functioned not merely as localization tools but as complex depictions that alluded to a cultural and social sense of territory, and could thus be effectively used to ground our research.

All that remained was to make them widely available, along with a suitable set of directions for their interpretation. That would be done via online resources and representation toolsets of *cybercartography*.

6 From GIS to reflexive mapping

In the first months of the contagion, persistent but cursory references to cartography made by the media to communicate or explain the spread of the epidemic led us to envision maps that, setting aside trivial localization functions assigned to them by the media, would raise topography to the level of *chorography:* a reflexive mapping able to assemble both epidemic and social data based on constructive cues from cartographic semiosis[v] (Casti, 2015).

This theory postulates that once the field has been cleared of neo-positivist assumptions whereby maps are objective and neutral tools, representations will play a crucial role not so much and not only as descriptors of the world but also as blueprints for modifying the world, thanks to the interpretative toolsets they provide.[w] In other words, maps are given a mediating role between reality and representation, whereby reality itself can be actively shaped.[x]

If we were to disregard this today and consider maps merely as one of the many modes of representations, which make up the image of the world, we would be grossly shortsighted. Representation obviously has its purpose, but its strength does not end in setting forth a sign system capable of conveying in an orderly manner what would otherwise appear too complex. Rather, what helps us to understand why representations are so essential in the relationship people establish with their world is the fact that the world actually takes shape through them. In short, a cognitive act is already a selection of attributes among possible others. However, this act is transformed and becomes interpretational when it is recognized as an instance of representation: a model that must necessarily rely on a sign system in order to express itself.

Therefore, as we enter the semiotic field to pursue this line of inquiry, what we need to stress is that every representation presupposes the adoption of a sign system capable of transmitting information. The next step is to become aware that not all representations must be placed on the same level with regard to their communicative effectiveness, or rather their ability to convey an orderly model of the world as the only one possible.

There is no doubt: Maps historically figure as the most effective instance of representation in this sense. Not only do they convey an orderly model but they also impose such a model by *iconizing it,* that is, to say, by setting it forth on the basis of an interpretative theory that varies according to the sign system on which a given map is built. In topography, for example, this

[v]The term "reflexive cartography" is meant to draw attention to the researcher/cartographer as a figure engaged in both the study and the solution of socially relevant issues, including the role played by GIS in *empowerment* or the potential cultural assimilation brought about by these tools.

[w]As Bourdieu convincingly argued, the effect of a representation is not based on its objectivity or its subjectivity, since both tend to establish what exists and what does not exist (Bourdieu, 1991).

[x]That sheds light on Dardel's conviction that objectivity is not in itself a guarantee of final truth and ought not be accepted without reserve, since modern individuals draw objectivity from their subjectivity (Dardel, 1986).

theory consists in measurement: Objects must be scaled to reality, are located at a precise distance from each other, and respond to abstract symbolization criteria that refer to quantity. All this guarantees an objective and neutral topographical representation, which is as reliable or possibly more reliable than reality itself.

Furthermore, maps seem endowed with another powerful feature: They can generate discourse, that is, they can produce self-referential information not foreseen by the cartographer. Maps have been shown to function semiotically on several communicative levels, so much so that they do not merely invest things with meaning but produce meanings *from scratch* (Casti, 2000, 2015). This cartographic feature, defined as *self-reference*, is the actual engine of cartographic communication. Although it is activated by an interpreter, communication here refers to the self-referential working of maps. To use a theatrical metaphor, we may consider an interpreter as an actor on the "communicative scene", which is directed by self-reference: It is the director who dictates the model whereby interpretation is to be achieved.

Thus, one conclusion seems inescapable: Interpretation cannot possibly be considered as a purely cognitive or, if you will, neutral operation. Rather, to interpret maps is to implement an instance of territorial action, to envisage phenomena that will take shape as they are read out on the map, and to become part and parcel of social knowledge. Think for instance of an interpreter acknowledging the existence of a phenomenon via a map: Such an interpreter simply accepts the map as a mediator and embraces the proposed form of the given phenomenon as the undisputed premise for his own actions. If a map charts a distribution of contagion based on a topographic metric, it records the phenomenon's relevance based on the metric extension of regions rather than on the number of individuals who inhabit those regions. On the other hand, if we consider inhabitants as the relevant data on the map and distort the dimensions of those areas by expanding or contracting them, we will "humanize" the contagion and show its actual societal import.

In short, once a semiotic perspective is embraced against the strict requirements of accuracy and objectivity typical of topographic metrics, new horizons open up for interpretation. More importantly, new models become available for building maps capable of inducing mediatization, that is, a diffusion and amplification of complex phenomena to be represented, finally shown on the basis of criteria no longer strictly tied to metric dimensioning.

It goes without saying that if this possibility is in fact pursued and the map outcome is adjusted, manipulated, or tweaked while the map is constructed, ethical precautions must be taken with regard to interpreters. In the first place, choices, techniques, and aims must be explicitly stated and made transparent. It should be stated clearly that maps will be adjusted to reflect a specific goal, namely, *not to provide an objective or neutral representation of the world, but to put forth a model of world phenomena grounded in their social relevance and complexity*.

To avoid feeding false expectations, we should make it clear that what is presented here is not meant to fully address the radical (but also practical) question of laying out new principles or providing prescriptions on how to set up a new cartography. Rather, our aim is to provide an overall view of contemporary experimentation in the realm of cartography: a variegated and complex scenario, and a harbinger of new perspectives.

This book also proposes a cartography that alludes to the social sense of territory by distorting the topographic map base, implementing it, and treating it via choices and techniques, which are dictated by "other" metrics: for instance, "chorographic" metrics, aimed at representing the social relevance of territory. We bring out the communicative potential of maps via specific tweaks or simple adjustments. In addition, we envision a long-term reflection on the potential adaptation of cartography to a societal view of the world. In short, as we inhibit topographic metrics and reclaim a topology of places, we set forth a model of chorographic metrics, achieved via digital technology, which allows for new interactions between cartographers and recipients and opens up new lines of study on the Covid-19 epidemic as well.

Familiarity with the constructive and communicative mechanisms at work in cartography makes it possible to have an effective recourse to digital and online tools for geographical information. However, the aim is less to use technology for a localized representation of phenomena and more to use their quantitative spatialization in order to convey the qualitative aspects of the contagion, namely, its propagation, intensity, and severity of infection in certain areas. Thus, we will be shifting communication from the depthless level of denotation to the highly meaningful level of connotation. The striking differences whereby contagion-affected regions may be investigated cartographically, via cross-referencing with differences that pertain to the physical, social and local environment. Accordingly, a contextualized framework on a local and national scale may be drawn. As it yields information about the relationship between contagion and socio-territorial systems, this data processing also brings to the surface the fragility of contemporary living that the pandemic has exposed. These are shortcomings tied to structural features that facilitate the contagion, such as pollution, or aspects related to our mobile and urbanized living, which favors contacts and crowding, for instance in commuting. They may also be flaws ascribable to the health-care system and lack of adequate local facilities. All these issues lay the ground for rethinking territorial policies both during and after the Covid-19 pandemic. Therefore, the role of *digital mapping* in this context is highly relevant, precisely because it implies reflexivity on what has happened and on what happens. Our understanding of the spread of contagion and of its local differences depends on that. Symbolic operator[y] maps ultimately point out specific features that call for urgent intervention against pandemics. They also encourage wider reflection and an informed questioning of the western model, by now globalized, of inhabiting the earth.

All this is made possible by the far-reaching technical potential of GISs: the large amount of data GISs can manage and the unlimited number of attributes for each geographical phenomenon; the possibility of processing and reclaiming spatial relations otherwise hard to detect; the ability to integrate different datasets, at different scales and from a wide range of sources; and finally, on the basis of the same data, the ability to draw representations that differ each time, thanks to the distinction between an archival function, entrusted to the *database*, and an iconic function, achieved in the process *output*. There is no room for doubt: on account of their vast array of resources, GISs enable us to move away from topographic metrics in favor of a chorographic model.

Certainly, a quantum leap was achieved when GISs were integrated with online technology. The paradigm shift occurs because GISs online are no longer final products but developing constructs: dynamic cartographic models were never finalized or thoroughly

---

[y]Maps may become *symbolic operators* when they are freed from their conventional role of mere recorders of reality and are seen instead as *media* of hypertextual communication, at once capable of describing and conceptualizing the world, of explaining how it functions on the basis of a theory, which also entails active intervention in territorial practices (Casti, 2015, pp. 17–19).

defined. Online GISs make it possible for anyone to make or unmake maps. There never is a finished product. Instead there will be an ever-changeable model, which will have to be re-thought less on the basis of technical requirements than on the needs of actual communication: a radically new tool that brings out the semiotic working of cartography while it sets forth its own semiotic modeling.

That is where *cybercartography* comes in: *chorographic* dynamics that relies on multimedia and shifting perspectives to represent a space–time dimension, which reclaims the social value of territory, in our specific case with regard to a viral contagion.[z]

This marks a momentous objective, since WebGIS[aa] applications exploit GIS software analysis and use classic *web-based* functionalities to reach a wide audience of expert or nonexpert users via multiple web platforms. When used competently, these online geographic information tools may help disseminate both the relevance of the relationship between socio-territorial aspects and pandemic contagion and the potential of new digital *mapping* systems.[ab]

7 Conclusions

The term *"cybercartography,"* in the sense laid out by Fraser Taylor (2006); Fraser Taylor et al. (2019), identifies digital maps aimed at reclaiming the social and cultural values of communities, which are conveyed through the communicative and pragmatic potential of digital cartography.[ac]

In this context, however, and with an eye on cartographic semiosis, *cybercartography* is proposed not only as an effective communication tool for representing spatial-temporal social phenomena such as Covid-19 but also possibly as a *symbolic operator*, that is, an operational tool for a social management of a contagion, which actively influences decision making.

[z] An instance of *digital mapping* on the Covid-19 epidemic in Italy may be found at https://cst.unibg.it/it/avvisi/mapping-the-epidemic-systemic-geography-of-covid-19-italy.

[aa] WebGIS partake of *online* cartography or the set of electronic maps widely available online. This category includes different types of representations both in terms of map processing and in terms of the resulting map formats offered to final recipients (Casti, 2015, pp. 142–145).

[ab] Covid-related cartography circulated at present arguably obfuscates the potential of digital cartographic systems because computer scientists, who are neither mapping analysts nor experts on territory, tend to entrust data representation entirely to computer algorithms, so that information is neither directly managed nor made thoroughly intelligible. We do have one study conducted by a group of researchers in China, which seems to pursue the cartographic processing outlined here: the use of logarithmic processing for cross-referencing epidemic data to population data. See Zhou et al. (2020).

[ac] Fraser Taylor introduced the concept of *Cybercartography* at the International Cartographic Conference in Sweden in 1997. He convincingly argued that if cartography was to play a more incisive role in the information age, a new paradigm was called for. Fraser's studies evolved in the wake of "postmodern cartography" as initiated by John B. Harley in 1989 and developed over the first decade of the 2000s by others. For a general survey, see Azócar Fernández and Buchroithner (2014). This new cartographic model aimed at organizing and communicating spatially related information on a wide variety of topics, which are relevant to society. *Cybercartography* relies on an interactive and dynamic format based on multimedia and aims at communicatively implementing a *chorography model*, which is meant to convey the social values of territory and the cultural values of landscape (Casti, 2015).

On the basis of these theoretical assumptions, citizens may be involved in the process of producing a digital map of contagion, and that may in turn branch out to become a model of territorial *governance*, an instance of interactive democracy that envisions maps as symbolic operators affecting decision making.

Thus, our "deck of cards" has been laid out: we can set out our strategy for facing the pandemic challenge on sound intellectual grounds.

References

Azócar Fernández, P.I., Buchroithner, M.F., 2014. Paradigms in Cartography: An Epistemological Review of the 20th and 21st Centuries. Springer, New York/Dordrecht/London.

Bourdieu, P., 1991. Language and Symbolic Power. Polity Press, Cambridge.

Casti, E., 2000. Reality as Representation. The Semiotics of Cartography and the Generation of Meaning. Bergamo University Press, Sestante, Bergamo.

Casti, E., 2015. Reflexive Cartography. A New Perspective on Mapping. Elsevier, Amsterdam.

Dardel, E., 1986. L'uomo e la terra, natura della realtà geografica. Unicopli, Milan.

Djaiz, D., 2020. La mondialisation malade des ses crisis? In: Le Grand Continent. accessed December 2020 from https://legrandcontinent.eu/fr/2020/03/23/coronavirus-mondialisation-david-djaiz/.

Fraser Taylor, D.R., 2006. The theory and practice of cybercartography: an introduction. In: Fraser Taylor, D.R., Lauriault, T. (Eds.), Cybercartography, Theory and Practice. Elsevier, Amsterdam, pp. 1–13.

Fraser Taylor, D.R., Anonby, E., Murasugi, K. (Eds.), 2019. Further Developments in the Theory and Practice of Cybercartography. Elsevier, Amsterdam.

Hall, P., Pain, K., 2006. The Polycentric Metropolis. Learning From Mega-City Regions in Europe. Earthscan, London.

Jackson, M.O., 2020. Comment se diffuse un virus? In: Le Grand Continent. accessed December 2020 from https://legrandcontinent.eu/fr/observatoire-coronavirus/.

Lévy, J., 2008. Un évènement géographique. In: Lévy, J. (Ed.), L'invention du monde. Une géographie de la mondialisation. Presses de Sciences Po, Paris, pp. 11–16.

Lévy, J., 2020. L'humanité habite le Covid-19. AOC. Analyse, Opinion, Critique. accessed December 2020 from https://aoc.media/analyse/2020/03/25/lhumanite-habite-le-covid-19/.

Lussault, M., 2007. L'Homme spatial. La construction sociale de l'espace humain. Seuil, Paris.

Lussault, M., 2017. Hyper-lieux. Les nouvelles géographies de la mondialisation. Seuil, Paris.

Lussault, M., 2020. Chroniques de géo' virale. Ecole urbaine de Lyon/Editions deux-cent-cinq, Lyon.

Maggioli, M., 2015. Dentro lo Spatial Turn: luogo e località, spazio e territorio. Semestrale di Studi e Ricerche di Geografia XXVII (2), 51–66.

Soja, E., 2000. Postmetropolis: Critical Studies of Cities and Regions. Blackwell Publisher Ltd., Oxford.

Warf, B., Arias, S. (Eds.), 2009. The Spatial Turn. Interdisciplinary Perspectives. Routledge, New York.

Zhou, C., et al., 2020. COVID-19: challenges to GIS with big data. Geogr. Sustain. 1 (1), 77–87.

Chapter 1. Population and contagion spread

1.1

Evolution of epidemic outcomes in Europe

Elisa Consolandi

1.1.1 Premise

Covid-19 is considered an ongoing global epidemic event that began in Wuhan, in the Hubei region of China, towards the end of 2019. The virus's ability to spread rapidly meant that contagion reached most territories on a global scale by March 2020, and qualified as a *pandemic* event, as stated by the World Health Organization (WHO).[a] To this day, the factors and agents that may have caused such worldwide contagion remain undefined. However, the paths that have facilitated its extremely rapid diffusion derive arguably from the dense network of movement and connectivity that globalization has brought about.[b] The pervasiveness of contacts between people, goods and information engenders a multi-scale phenomenon which involves simultaneously two globalized dimensions (local and global). This suggests that viral spreads occur both by proximity, i.e., implicating proximity as a spatial factor, and by reticularity, as a result of the multiple connections, which depend on the dynamic interaction of territories.[c] As we chart epidemic evolution cartographically, we intend to claim the relevance of these spatial components for setting up a line of research that, going beyond the boundaries of biomedical and epidemiological disciplines and employing general geographical and social skills, addresses the complexity of the Covid-19 phenomenon from an interdisciplinary perspective.

[a]In a press conference held on 11 March 2020, WHO Director general announced that Covid-19 was to be characterized as a pandemic. The press conference transcript may be found here: https://www.who.int/dg/speeches/detail/who-director-general-s-opening-remarks-at-the-media-briefing-on-covid-19--11-march-2020.

[b]On the network dynamics created by globalization, see among others: Lévy (2008).

[c]The analysis we envision in this section embraces the theoretical approach and the methodology proposed by Emanuela Casti and laid out in the introduction to this volume, according to which the spread of Covid-19 spread may be assessed to have taken place both by proximity and by reticularity. For a full treatment of this issue, see also: Casti (2020).

19

1.1.2 Covid-19 in Europe

The spread of Covid-19 in Europe began with the identification—across several countries—of isolated cases attributable to severe acute respiratory syndrome coronavirus 2 (SARS-CoV-2), followed by viral outbreaks over the first 2 months (January–February) of 2020. Ever since the first Covid-19 cases were recorded in Europe, supranational institutions were mobilized to coordinate and manage the emergency through the provision of care resources and protective equipment, as well as via the collection of data and information on the spread of the virus, on actions to be taken in order to contain it and on the measures to be adopted for alleviating the economic and social damage caused by the pandemic.[d] All this was not enough, however, to stop the devastating impact of Covid-19 in Europe, which—to this date (June 2021)—has exceeded 54 million cases and has led to over 1 million deaths.

The first visualization for Europe (Fig. 1.1) is in the form of a mosaic map which displays the absolute number of outbreaks and their distribution across continental Europe in four distinct phases between March and June 2020. These phases were selected with regard to their temporal frequency, which makes it possible to chart their evolution notwithstanding the

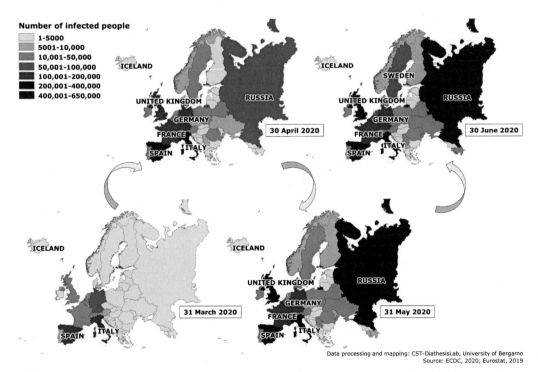

Data processing and mapping: CST-DiathesisLab, University of Bergamo
Source: ECDC, 2020; Eurostat, 2019

FIG. 1.1   Europe: national distribution of Covid-19 infection from 24 February to 30 June 2020 (absolute data).

---

[d] These initiatives are laid out on the following webpage of the European Council, which outlines the response to Covid-19 of the European Union in the public health sector: https://www.consilium.europa.eu/en/policies/coronavirus/covid-19-public-health/ (last accessed: 24 February 2021) and are analyzed in Chapter 6 of this volume.

measures progressively adopted to combat contagion in the different countries. It should be noted that this selection was made given the impossibility of establishing the actual efficacy of containment measures in the various countries, since legal measures prove effective only when enforced. Cultural factors, media communication perspectives, public trust, and an acknowledgement of the authoritativeness—as well as the authority—of public institutions in effectively pursuing the common good: all these factors determine a variegated pattern in the relation between regulatory provisions and social outcomes which cannot possibly be recalled in the present context. Hence, a comparison between the various epidemic phases in which the virus circulated is based on cadenced sampling of the period and can only record the quantitative spatial–temporal spread of contagion. As of March 31, 2020[e] the virus had reached Western Europe, that is the most connected and globalized European territory. In absolute terms, contagion affected first of all Italy and Spain, followed by Germany in number of infections with France, the United Kingdom, Austria, the Netherlands, and Belgium trailing behind. By April, Covid-19 contagion gradually spread throughout Eastern Europe, where it affected Russia in particular before growing progressively severe across Belarus, Poland, Romania, and Sweden. In addition, significant escalation of the Covid-19 epidemic may be noted across Western Europe, specifically, Spain and Italy—followed by Germany, France, and England— where contagion grew progressively worse and rapidly reached high absolute numbers. Such data highlight the remarkable speed of viral contagion within a limited time frame.

A subsequent sampling records an escalation of contagion following a gradual worsening of the epidemic across the whole European area which nonetheless seems to have spared—at least in absolute terms—Iceland, Latvia, Estonia, Lithuania, and the states belonging to the western Balkan Peninsula. Such expansion of the disease marks an increase in the number of people suffering from SARS-CoV-2, which reaches very high figures in some states (specifically: Russia, United Kingdom, Spain, and Italy, with Germany and France trailing shortly behind). Analysis of contagion from a spatio-temporal perspective relies on Covid-19 localization to render the devastating force whereby the epidemic swept Europe and to identify the areas most affected by the infection. Such data interpretation is made possible exclusively by a spatialization of information and by tracing the quantitative temporal development of contagion.

We might at this point forgo a conventional model of thematic cartography which uses spatiality, i.e., the extent of areas, to make sense of data, and turn instead to a model of reflexive cartography which cross-references such data to other related datasets and ensures a more articulate understanding of the map's social implications.[f] For example, if infection data are compared to the resident population of each country,[g] the given information outlines

---

[e]Contagion data for the European scale are available starting from the end of February 2020, but not for all countries, whose data were systematically recorded starting from the first days of March. Accordingly, dates for the space–time analysis were selected at regular intervals with a view to charting the epidemic course gradually.

[f]This chapter embraces the theory of cartographic semiosis laid out by Casti (2000) and her concept of "reflexive cartography" (Casti, 2015). The idea is to promote representations able to integrate multiple datasets, by using a space–time approach and privileging 3D techniques and anamorphic visualization. For more details on the reflexive approach see: Casti (2015).

[g]An extended assessment of urban and non-urban populations residing across Europe would have been desirable but falls outside the limits of this research, whose main focus is Italy.

trends or specifies their social relevance. Reflexive cartography relies on graphic tools, such as tridimensional representation[h] (Casti, 2015, pp. 165–168) of state areas, to point out that a low number of infections matches in fact a high contagion incidence, a datum which underscores the social dimension of viral spread. Infographic processing is meant here to raise awareness of the fact that *data in themselves do not provide information* until processed statistically and graphically localized in a mapping model that cross-references multiple datasets.[i] Statistics-based data processing makes it possible to compute a contagion index by country and hence estimate the impact of Covid-19 on resident populations. However, it is only via reflexive cartography that such data can be properly appraised in a comparison across states, both because the latter are localized and because their demographic features are explicitly retrieved.

So, for instance, the (Fig. 1.2) Covid-19 distribution map as of June 30, 2020, visualizes both an absolute number and a contagion index in Europe. Extrusion here endows state areas with a double role, namely recording the quantity of infected people and at the same time showing their percentage incidence over the total resident population. The absolute number of infected people is rendered through vertical extrusion of state areas, while color-coding conveys the incidence of infection on the population. Geo-visualization of the ratio between color-coded and extruded areas provides a detailed and comprehensive assessment of the social impact of Covid-19. For example, high Covid-19 incidence in Iceland is color-coded in red. However limited extrusion of Iceland's geographical area indicates a low number of infections.[j] The ratio between the two data underlines that high incidence in this case derives from the island's low population density.[k]

Russia stands out in the map given the expanse of its territory, which via extrusion conveys the large number of infections. Viral incidence is also considerable, but well within range, as

[h]The technique of *extrusion* makes it possible to visualize the intensity of contagion and to underline its severity in absolute numbers, which are represented by 3D polygons. At the same time, information about the incidence of infection may be derived from color-coding and the cross-referencing of data on infected people with data on residents for each country.

[i]In a volume on *Basic Epidemiology*, WHO researchers touched upon the issue of raw epidemic data as information that is undoubtedly useful but calls for further interpretation if it is to be read effectively. It was noted that "the number of cases alone without reference to the population at risk can occasionally give an impression of the overall magnitude of a health problem, or of short-terms trends in population, for instance during an epidemic. WHO's *Weekly epidemiological record* contains incidence data in the form of case numbers, which, in spite of their crude nature, can give useful information about the developments of epidemics of communicable diseases" (Beaglehole et al., 1993, p. 14).

[j]With just over 350,000 residents (as of 2019), Iceland is one of the least densely populated countries in the world. Most people reside near the urban area of the capital Reykjavík, since much of Iceland's land is taken by mountains, plateaus, and a substantial number of glaciers. Land morphology certainly favored distance between settlements, which nevertheless record a dense network of relations.

[k]Similarly to Iceland, although not visible on this map, the Republic of San Marino, Vatican City, as well as Andorra and Gibraltar all report a low number of infected people within their territories yet show a very high incidence of Covid-19 infection. Such countries are among the least populous on the European continent.

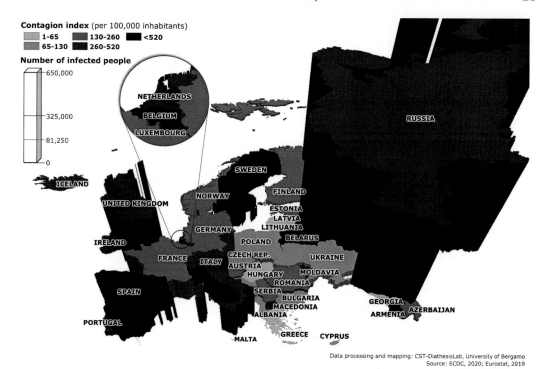

FIG. 1.2 Europe: distribution of Covid-19 infection as of 30 June 2020.

attested by color-coding. Also in this case, reflexive mapping mitigates the possible alarm absolute infection numbers might cause: the ratio between infection numbers and incidence data redresses the balance with respect to other countries.[1]

Another example is that of Spain whose significance, however, emerges when compared with other countries in Western Europe. Dark red color-coding conveys Spain's high viral spread index, while marked extrusion indicates an absolute high number of infected people. The combination of color-coding and extrusion provides an effective and immediate representation of the devastating social outcome of the healthcare emergency in Spain. Once again, if taken in isolation, absolute data on infected individuals do not allow effective comparison between states, while incidence relative to a fixed population number do. Analysis may also be extended to the United Kingdom and Italy, whose territories as of June 30, show a high level of contagion, which is rendered both through a tridimensional geo-mapping of territory and by color-coding. Finally, Sweden, Belarus, Luxembourg, Belgium, and Armenia deserve a separate mention. While these countries do not record high percentages of absolute

---

[1]It should be remembered that the Russian Federation, with its 145,872,260 inhabitants, is the most populous territory on the European continent (See:https://www.ecdc.europa.eu/en/publications-data/sources-worldwide-data-covid-19).

contagion data, as visually suggested by extrusion,[m] they do show high incidence rates, color-coded in darker shades. As seen in other cases, such ratio may be ascribed to lower population density than in the rest of Europe.

1.1.3 Factors favoring viral propagation

At least in the first phase of the epidemic, atmospheric pollution and particulate emissions may count as two major co-factors facilitating the spread of Covid-19.[n] Observation and analysis of the concentration of toxic elements in the atmosphere relative to viral spread on a European scale has led researchers to zero in pollution levels around more urbanized areas, where contagion has been most severe.[o] Atmospheric pollutants are assessed as risk factors, since they may cause or facilitate cardiovascular disease and respiratory viral infections.[p] The onset of these pathologies seems in fact to be favored or triggered by prolonged exposure to atmospheric pollution, in particular nitrogen dioxide (NO_2), a toxic byproduct of combustion in energy production and consumption for the domestic and industrial sectors.[q] Nitrogen dioxide, as well as microparticulate (PM_X), is among the atmospheric pollutants that may have led to a greater spread of SARS-CoV-2,

---

[m] As of June 30, the absolute number of infected people in Sweden was 67,667; in Belarus it equaled 49,609; and in Armenia there were 25,127 people affected by Covid-19. On the same date, incidence of the number of infected people in Sweden was 661.44 per 100,000 inhabitants; in Belarus it was 524.83 and in Armenia it was 849.54. Although extrusion distortion makes the two areas invisible on the map, it is important to underline that the Grand Duchy of Luxembourg had an incidence of infection equal to 693.28, while Belgium recorded 537.49 people affected by the virus per 100,000 inhabitants.

[n] Research papers on this issue in Italy and abroad are numerous. From different points of view, they address the correlation between viral spread and the massive presence of atmospheric pollutants, taken as risk factors that may favor respiratory tract infections. Among these: for environmental health sciences see Benmarhnia (2020); Conticini et al. (2020); with regard to atmospheric and climate sciences, see Contini and Costabile (2020); as regards environmental epidemiology, Cori and Bianchi (2020), Filippini et al. (2020); finally, for remote survey and cartographic production in relation to pollution data, consider Ogen (2020); Pansini and Fornacca (2020).

[o] In addition, studies have been published with the aim of monitoring changes detected in air quality and tracing the progressive reduction of polluting emissions due to the strict containment measures (*lockdowns* in particular) which were adopted to avoid viral spread across European regions. The ESA-European Space Agency website records a decrease in air pollution levels in the months of March–June 2020 compared to the same period for the previous year (www.esa.int).

[p] Close linkages between cardio-vascular pathologies and pollutants have also been highlighted in recent international research focused on the spread of Covid-19, such as: Barcelo (2020).

[q] Long-term exposure to high concentrations of nitrogen dioxide can lead to an increased risk of asthma and lung dysfunction, as well as a significant increase in mortality from cardiovascular and respiratory distress. Furthermore, for an in-depth analysis of atmospheric quality levels across the European territory, you may consult the *report* published by the EEA-European Environment Agency (EEA, 2019).

since it is one of the major factors affecting the immunological health of individuals (Conticini et al., 2020), especially when subjects are weak or predisposed to pathogen invasion and viral diseases.[r]

While research is still ongoing, initial analysis of the spread of Covid-19 relative to the level of NO$_2$ recorded in the last weeks of February 2020—the period in which outbreaks were first detected on European territory—suggests a correspondence between major urbanized areas of the countries where contagion was most severe like Northern Italy,[s] and the level of pollution present in the troposphere (Ogen, 2020). Air pollution surveys carried out by leading space agencies in Europe[t] or the US (ESA; NASA)[u] at the onset of Covid-19 contagion in Europe show a higher concentration and spread of nitrogen dioxide in some regions of the continent. Specifically, pollutants were recorded in some areas that from Great Britain extend to Germany and reach all the way east to Moscow and south to Italy. The areas most affected by the presence of atmospheric pollutants in Europe are: the Po Valley[v] in Northern Italy; England (especially the London metropolitan area and in the cities of Leeds and Manchester); Île-de-France (hosting the Paris area); some areas of central Spain (the city of Madrid especially); parts of Poland (notably the Krakow area) and of Russia (specifically the Moscow metropolitan area). Air pollution extends across the territories of Belgium, the Netherlands and some areas of West Germany.

Other factors that contribute to viral propagation have to do with contemporary living patterns, to be traced specifically to commuting practices and gatherings that occur in public spaces. In this sense, representations involving Europe identify a backbone that extends between Lombardy and the United Kingdom (Fig. 1.3) and includes densely populated areas,

[r]Even though research in this field has yet to achieve systematization, a number of studies—such as those proposed by the Italian Society of Environmental Medicine (SIMA)—have advanced the hypothesis that atmospheric pollution, in particular PM$_X$ and NO$_2$ particles suspended in the air, may have played a key role in the propagation of Covid-19, as vectors (or *carriers*) responsible for viral transmission via aerosol (Setti et al., 2020a,b). It is well known that the primary cause of airborne infection are breath droplets inhaled by proximity to infected individuals.

[s]Northern Italy and the Po Valley in particular is a highly industrialized area and one of the most polluted in Europe—as evidenced by the recent report of the EEA-European Environment Agency (EEA, 2019). Its geographical layout favors air stagnation and, consequently, the persistence of pollutants (ISTAT, 2020, p. 45).

[t]The online database on air pollution by nitrogen dioxide that we consulted is based on information provided by the Copernicus Sentinel-5P satellite, designed to perform high space–time resolution atmospheric measurements, with a view to detecting and monitoring air quality and air pollutants (see: https://www.copernicus.eu/en, https://maps.s5p-pal.com/).

[u]The scientific team involved in NASA's Earth Observing System develops and validates algorithms through satellite instruments which offer a global view of the planet's atmospheric pollution. The aim is to trace the intensity of nitrogen dioxide present in the troposphere. Satellite measurements of pollutants have proved invaluable for air quality studies. For more details see: https://so2.gsfc.nasa.gov/no2/pix/regionals/Europe/Europe.html.

[v]For more details, you may see Chapter 4 of this volume and specifically Fig. 4.1.

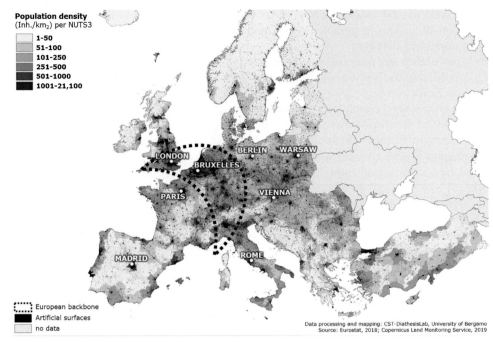

FIG. 1.3 Distribution of urban areas and population density in Europe.

areas of trade and large traffic cross-border.[w] Although Covid-19 contagion cannot be fully ascribed to the population density of one territory alone,[x] it should be underlined that, in most metropolitan areas, hyper-places and public spaces, which lead to gatherings of people, strong commuting flows and high mobility have favored the transmission of viral diseases, since "closer contact and more interaction among residents [...] makes them potential hotspots for the rapid spread of emerging infectious diseases." (Hamidi et al., 2020, p. 2). In this context, social distancing and the subsequent restriction of movements played a key role in containing outbreaks, since it limited personal interactions and gatherings which made it possible for Covid-19 to reach the whole urbanized world via multiple networks. The viral spread is undoubtedly favored by interactions and mobility between inhabitants, which

[w] A line of research pursued at the Observatoire Géopolitique du Covid-19 envisions the spread of infection as extending along the area of the European backbone first identified by Brunet (1989) in a study that was conducted under his supervision at the Reclus institute of Montpellier in France. The backbone refers to an urban corridor which comprises a large number of small and medium-size cities, from Northern Italy to the south-eastern part of Great Britain. Such cities are closely linked to large metropolitan regions, considered by French researchers as the driving forces of European socio-economic development. The research group at the Observatoire Géopolitique du Covid-19 has been monitoring the trend of infections in European territories for months. Research results are published in an online journal entitled *Le Grand Continent*; see their web page: https://legrandcontinent.eu/fr/observatoire-coronavirus/ (last accessed: 4 October 2020).

[x] The relationship between population density and viral spread has long been a matter of debate, as outlined by: Hamidi et al. (2020).

extend well beyond their immediate vicinity and reach distant places also by virtue of major commuter and trade flows. This leads us back to the claim of French geographer Jacques Lévy, whereby the spread of viruses must not be seen exclusively as a biological phenomenon, but as a phenomenon necessarily influenced by a social component (Lévy, 2020).

1.1.4 Conclusions

The present analysis set out to provide an overall picture of the spread of SARS-CoV-2 across the European territory during the first wave of infection. The aim was to draw a synoptic chart of the epidemic emergency on the basis of territorial differences, in order to verify the hypothesis which underlies this study, namely that territorial factors affect the modes of Covid-19 contagion. In Europe, Covid-19 severely affected the most connected and globalized Western countries first, and notably Italy, in terms of intensity and gravity. Later, contagion reached member states of the easternmost European continent. While the processing of data on the Covid-19 epidemic collected between the months of February and June 2020 does not provide a thorough picture of the dynamics of contagion on a European scale, it does detect contagion trends that seem to relate to a space–time model of data interpretation. In addition, analysis of physical and social factors corroborates the hypothesis that contagion did not strike epicenter areas arbitrarily but was strongly correlated to and possibly dependent on territorial features.

In this context, cartography is an invaluable tool for rendering the complexity of the ongoing Covid-19 epidemic, through the aid of reflexive maps and the application of innovative graphics for territorial representation. Reflexive geo-visualization underscores the relevance of territorial features for a sharper and more articulate assessment of the epidemic. The processing of multiple data—both referred to contagion and inherent in socio-territorial features—make it possible to improve our understanding of the dynamics triggered by Covid, and to propose useful interpretative models for coping effectively with the emergency.

References

Barcelo, D., 2020. An environmental and health perspective for Covid-19 outbreak: meteorology and air quality influence, sewage epidemiology indicator, hospitals disinfection, drug therapies and recommendations. J. Environ. Chem. Eng. 8 (4), 1–4.

Beaglehole, R., Bonita, R., Kjellström, T., 1993. Basic Epidemiology. World Health Organization, Geneva.

Benmarhnia, T., 2020. Linkages between air pollution and the health burden from COVID-19: methodological challenges and opportunities. Am. J. Epidemiol. https://doi.org/10.1093/aje/kwaa148.

Brunet, R., 1989. Les Villes européennes, Rapport pour la DATAR. La Documentation Française, Paris.

Casti, E., 2000. Reality as Representation. The Semiotics of Cartography and the Generation of Meaning. Bergamo University Press, Sestante, Bergamo.

Casti, E., 2015. Reflexive Cartography. A New Perspective on Mapping. Elsevier, Amsterdam.

Casti, E., 2020. Geografia a 'vele spiegate': analisi territoriale e mapping riflessivo sul Covid-19 in Italia. Documenti Geografici 1, 61–83.

Conticini, E., Frediani, B., Caro, D., 2020. Can atmospheric pollution be considered a co-factor in extremely high level of SARS-CoV.2 lethality in northern Italy? Environ. Pollut. 261, 1–3.

Contini, D., Costabile, F., 2020. Does air pollution influence Covid-19 outbreaks? Atmos. 11 (4), 377–381.

Cori, L., Bianchi, F., 2020. Covid-19 and air pollution: communicating the results of geographic correlation studies. Epidemiol. Prev. 44 (2–3), 120–123.

EEA, 2019. Air Quality in Europe—2019 Report. Publications Office of the European Union, Luxembourg.

Filippini, T., Rothman, K.J., Goffi, A., Ferrari, F., Maffeis, G., Orsini, N., Vinceti, M., 2020. Satellite-detected tropospheric nitrogen dioxide and spread of SARS-CoV-2 infection in Northern Italy. Sci. Total Environ. 739, 1–7.

Hamidi, S., Sabouri, S., Ewing, R., 2020. Does density aggravate the COVID-19 pandemic? J. Am. Plann. Assoc. https://doi.org/10.1080/01944363.2020.1777891.

ISTAT, 2020. L'inquinamento atmosferico. Rapporto sul territorio 2020. Ambiente, economia e società. accessed October 2020 from https://www.istat.it/storage/rapporti-tematici/territorio2020/Rapportoterritorio2020.pdf.

Lévy, J., 2008. Un évènement géographique. In: Lévy, J. (Ed.), L'invention du monde. Une géographie de la mondialisation. Presses de Sciences Po, Paris, pp. 11–16.

Lévy, J., 2020. L'humanité habite le Covid-19. AOC. Analyse, Opinion, Critique. accessed October 2020 from https://aoc.media/analyse/2020/03/25/lhumanite-habite-le-covid-19/.

Ogen, Y., 2020. Assessing nitrogen dioxide (NO_2) levels as a contributing factor to coronavirus (Covid-19) fatality. Sci. Total Environ. 726.

Pansini, R., Fornacca, D., 2020. Higher virulence of COVID-19 in the air-polluted regions of eight severely affected countries. medRxiv. https://doi.org/10.1101/2020.04.30.20086496.

Setti, L., et al., 2020a. Searching for SARS-CoV-2 on particulate matter: a possible early indicator of Covid-19 epidemic recurrence. Int. J. Environ. Res. Public Health 17, 2986–2990.

Setti, L., et al., 2020b. Evaluation of the Potential Relationship Between Particulate Matter (PM) Pollution and Covid-19 Infection Spread in Italy. Position Paper, S.I.M.A. accessed October 2020 from http://www.simaonlus.it/wpsima/wp-content/uploads/2020/03/COVID_19_position-paper_ENG.pdf.

CHAPTER

1.2

Italy into three parts: The space–time spread of contagion

Emanuela Casti and Elisa Consolandi

1.2.1 Introduction

In Europe, the Covid-19 epidemic affected Italy first: most notably Italian regions in the north. Subsequently, contagion also spread to the central and southern provinces of the Peninsula, but its diffusion was mild, and never reached the severity experienced in the north.[a] This resulting epidemic snapshot is complicated, but outlines the presence of "Three Italies," set apart in relation to the intensity and severity of Covid-19 infection. These three distinct areas have remained virtually unchanged in their geographical extension, but have grown progressively worse in terms of contagion. It was therefore necessary to define the territorial factors and specificities that may have determined this geographic setup, by looking both at climatic-morphological and socio-territorial features. Features of this kind in Italy are numerous and varied, so much so that they give rise to quite dissimilar territories both from a geophysical and a social point of view. In order to examine the reasons that may have led to such diversified spread of Covid-19 contagion across the areas considered, our analysis heavily relied on reflexive cartography, which allows effective cross-referencing of contagion data with socio-territorial information sets and may thus be envisioned as a privileged representation tool for charting the complexity of the epidemic. It should be

---

[a]This geographic model applies to the first wave of the epidemic in Italy, the one between February and June 2020: it therefore differs from the second wave, which began in October 2020. In this second wave, Italy seemed to be experiencing a more homogeneous spread of contagion, which has affected the entire Peninsula, with peaks in large cities (Rome, Naples, Florence, and Palermo) and whole tourist regions such as Sardinia or Trentino-Alto Adige. In autumn 2020, the lack of appropriate measure for addressing the crowding issue, often raised by commuting as people began to return to work, had a severe impact on large cities. At the same, internal tourism played a critical role in spreading Covid-19 to regions which had largely been spared in the first wave. See: Higher Health Institute (https://www.epicentro.iss.it/en/coronavirus/sars-cov-2-dashboard).

https://doi.org/10.1016/B978-0-323-91061-3.00012-0

underlined from the start that the maps we put forth and analyze below are not meant merely as visual tools for charting or monitoring Covid-19 outbreaks in Italy. Rather, to the extent that they keep track of the epidemic's space–time sequence, such cartographies also convey geophysical and social features as possible causes of contagion.

1.2.2 Mapping contagion: The spread of Covid-19 across the three Italies

The first Covid-19 epidemic wave (February–June 2020) struck suddenly and unexpectedly some northern Italian regions. This prompted researchers to look into a possible range of causes that may have facilitated such swift and far-reaching viral spread only in some areas of the Peninsula, while sparing others.

As discussed in the previous section, while at the European level, it was possible to trace the progressive spread of Covid-19 contagion in its first phase, across highly urbanized and interconnected areas of the most industrialized member countries, in the case of Italy multiple overlapping factors seem to apply and suggest that research ought to address more closely the physical and social setup of the Peninsula and its regions. Propagation of the SARS-CoV-2 virus[b] in Italy has brought out such significant variations—in terms of diffusion, intensity, and severity—that a tripartite model of the Italian national territory on the basis of the contagion seems in order, as indicated in Fig. 1.4 (Casti, 2020).

Cartographic representation charts the evolution of contagion in Italy at three specific times: initial outbreak; phase of maximum spread; and decrease phase. It employs the technique of anamorphosis[c] to distort territorial surfaces relative to the number of infected people, thereby highlighting the health crisis. Anamorphic distortion was implemented on provinces, that is on administrative districts within the Italian regions, which allows for a greater degree of detail in terms of contagion intensity.[d] Additional graphical processing was applied to contagion intensity data, which were color-coded along a gradient range of red. Each representation of Italy in Fig. 1.4 should thus be read as a mosaic, whose tiles are either expanded or contracted based on the number of patients residing in each province, and color-coded to rank data within a scale of values. Province by province, data processing for each map of Italy yields a quantification of the absolute number of infected people in a given phase and sets up a territorial ranking. Conversely, comparison between the three maps

[b] Data provided by the Italian Department of Civil Protection-Ministry of Health concern the total number of infected persons by region (NUTS 2) and by province (NUTS 3) recorded from 24 February 2020 (https://github.com/pcm-dpc/COVID-19).

[c] Anamorphosis is a representation technique that "undermines the assumptions of the topographic basemap by distorting them to fit the social data being represented [...]—the basemap is no longer the rigid container on which data are inserted, but a flexible, adjustable space that is forced, expanded, or contracted in relation to the relevance of the data" (Casti, 2015, p. 249–250).

[d] In fact, one of shortcomings of existing cartography on contagion which many national institutes have made available is a chronic lack of detail due, for instance, to the absence of appropriate map keys and the lack of designators for accurately spatializing data and color-codes without prior knowledge. This clearly impairs effective data communication, and possibly paves the way to misinterpretation and scaremongering. See: Casti (2020); Consolandi and Rodeschini (2020).

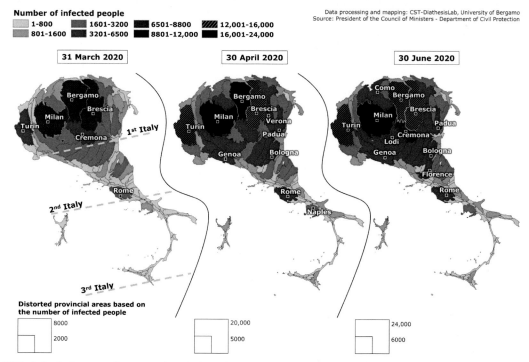

Number of infected people

1-800	1601-3200
801-1600	3201-6500
6501-8800	12,001-16,000
8801-12,000	16,001-24,000

Data processing and mapping: CST-DiathesisLab, University of Bergamo
Source: President of the Council of Ministers - Department of Civil Protection

FIG. 1.4 Italy: quantification and evolution of Covid-19 infection (absolute data).

makes it possible, on the one hand, to follow the temporal evolution of infection and, on the other hand, to evaluate its cumulative significance at the end of spring 2020, when infection in Italy entered an endemic phase.

As we turn to the first map about the situation at the end of March, we detect three distinct epidemic trends: Northern Italy, with a high number of infections that see Lombardy at the center; Central Italy, with sustained contagion only in Rome and in some provinces of the Marche region; and finally, Southern Italy, with islands recording only mild outbreaks.

The second map shows an intensification of contagion in Northern Italy, but that does not extend to the whole national territory. From Lombardy, the epidemic spreads to some provinces of Piedmont (Turin, in particular), of Emilia-Romagna, Veneto (with the cities of Verona and Padua recording the absolute highest data in the region), and Liguria. This trend is negligible in the rest of Italy, except for an intensification of outbreaks in some province cities of the Marche region (Pesaro-Urbino and Ancona[e]), as well as in the urban areas of Genoa, Florence, Rome, and Naples. Conversely, epidemic spread in the remaining part of the Italian

---

[e]These are provinces where the manufacturing industry is highly developed, especially as regards foreign trade, which sees Europe (notable Germany) as a natural market for Marche exports, followed by North America and Eastern Asia (namely, China). See: the Marche Chamber of Commerce (https://www.marche.camcom.it/).

peninsula, especially the islands and inland areas, is decidedly more contained, and surfaces for these areas on the map are anamorphically shrunk accordingly.

The third map refers to the endemic phase, which covers indicatively the months of May and June 2020, when Italy witnessed a decrease in the number of infections and the end of the national lockdown period. Contagion appears to have slowed down then, even though this map does not reflect that, because it takes stock solely of the total number of infections, highlighting regional disparities and confirming our tripartite contagion model of "three Italies" (*tre Italie*).

In summary, diachronic analysis of contagion (for March–June 2020),[f] on absolute data in Italy indicates that in the period considered, viral propagation intensified significantly in areas that were most affected initially, but the same north/south ratio was maintained throughout over the months. The Po Valley emerged as the Italian epicenter of Covid-19 epidemic, since it recorded the highest absolute contagion numbers, to which we should add figures relating to high mortality and outbreak severity.[g] That is in sharp contrast to the rest of the peninsula, which recorded an increasing spread of contagion over time, but with less intense and less severe outcomes.

As we turn from absolute data to an analysis of the contagion rate, the tripartite model of *tre Italie* is confirmed,[h] even though the disease epicenter has now shifted slightly, as shown in the first map in Fig. 1.5. The contagion rate is statistically obtained by comparing the number of infected people with that of the resident population. It enables researchers to assess contagion risk and is thus used as an indicator of the probability of a person becoming ill in a given territory. Once cartographically rendered, the contagion rate also makes it possible to estimate the level of danger for a given area in comparison to others.[i] Fig. 1.5, realized via the same procedure as the previous map, is most significant in this latter regard. While Fig. 1.4, on the basis of absolute numbers, indicated Milan, Bergamo, and Brescia, as the provinces initially most affected, Fig. 1.5 shows that the highest contagion rate is found instead in the neighboring provinces of the Po plain, namely those of Lodi and Cremona.[j]

---

[f]Italian Office for National Statistics data, referring to resident population in each province, date back to 2019 (https://www.istat.it/en/). Anamorphic mapping was obtained through shapefiles processing via ScapeToad. Province surface distortion reflects the number of infections in absolute terms. Color-coding gradients, once again referred to the number of people infected with Covid-19, were set and implemented through QGIS.

[g]On this issue, see Chapter 2 of this volume.

[h]The contagion rate was calculated by dividing the number of people infected with the virus by the number of residents in each province. The result was then multiplied by 100,000, to establish the number of Covid-19 cases recorded for every 100,000 inhabitants.

[i]Data addressed in this study refer to infection surveys for Italy carried out between February and June 2020. It should be noted that, after a period in which viral spread was generally contained and controlled, Covid-19 has regained strength over the following months, leading to a second wave of infections in Italy in October 2020.

[j]One should not forget that the province of Lodi was the very first area of Lombardy (and Italy) where a Covid-19 outbreak was recorded. Activity in this area was immediately curtailed by the Italian government, establishing the Lodi area as a "red zone". On the surge of outbreaks in Italy, see Chapter 5.1 of this volume. For more details on the Covid-19 Government Ordinance, see the decree issued in Italy by the President of the Council of Ministers on 1 March 2020 under the title: *Further operational provisions of decree-law dated 23 February 2020, n. 6, containing urgent measures for the containment and management of Covid-19 epidemic emergency.*

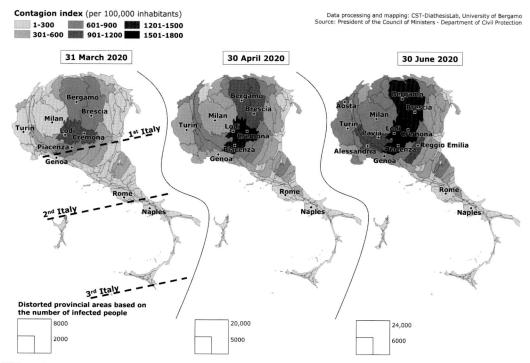

FIG. 1.5   Evolution of the Covid-19 contagion index in Italy.

It is only in the following month (April) that this rate also extends to the provinces of Bergamo and Brescia in Lombardy; those for which Fig. 1.4 had recorded the highest absolute number of infections. The province of Piacenza in Emilia-Romagna is added to these at that point.[k]

Overall, Fig. 1.5 also shows a gradual increase in the incidence of contagion over time in the north of the country,[l] replicating the distribution for absolute numbers outlined in Fig. 1.4. This corroborates the three Italies model we established earlier, with the Center, the South and the islands consistently showing a relatively low infection rate over the months, unlike Northern Italy.

No matter how the data are processed, they consistently suggest that since its inception the Covid-19 epidemic in Italy has spread along a tripartite geographical division. This subdivision is found to persist over time, as the three Italies model intensifies diachronically: a pronounced aggravation of infection for the first Italy is up against a moderate diffusion trend in

---

[k]In the metropolitan areas of Rome and Naples, contagion is higher in absolute terms, but not in terms of incidence rate among the population.

[l]This growth is particularly evident on 30 June 2020, when the provinces of Cremona, Lodi, and Piacenza— followed by the territories of Bergamo and Brescia—record the highest contagion rate in Italy.

the second and a virtually non-existent viral spread for the rest of the national territory. Ultimately, the most severe epidemic outcomes[m] were detected in the areas that were affected first, i.e., in the Po Valley, where Covid-19 had a lasting reach and a devastating impact also in terms of mortality.[n]

1.2.3 Morpho-climatic and socio-territorial factors

Marked differences in the intensity and severity of Covid-19 infection in the Italian regions have strengthened the hypothesis that multiple factors of a territorial nature, physical and/or social, may have influenced the spread of the SARS-CoV-2 virus. Such factors—such as, for example, the morphological and climatic setup of territory[o] or the type and density of urbanized areas inside it—point to highly relevant territorial features for understanding territorial vulnerabilities.[p] While such factors obviously fall short of directly explaining why the Covid-19 reaches dramatic proportions in Bergamo and precisely in Lombardy,[q] they do yield relevant suggestions for shedding light on the spread of contagion or on possible high-risk

[m] It was generally found that, within the time frame considered, Lombardy and Piedmont in particular recorded some among the highest percentages of hospitalized patients with symptoms, as well as a high number of infected people under home isolation. However, according to data provided by the Italian Department of Civil Protection-Ministry of Health, the percentage of patients who recovered is much higher than those who died due to the disease (http://opendatadpc.maps.arcgis.com/apps/opsdashboard/index. html#/b0c68bce2cce478eaac82fe38d4138b1).

[n] Conversely, as already suggested in note a, viral expansion over the entire Peninsula occurred in the second wave, in autumn 2020. Lombardy nonetheless remained the most affected region.

[o] Along these lines, special attention should be drawn to the level of atmospheric pollution recorded on the national territory and to the persistence of polluting ground emissions. For an in-depth treatment of pollutants see, among others: Murgante et al. (2020); Ogen (2020); Zhou et al. (2020). For a comprehensive treatment of this phenomenon in the Lombardy region, see chapter four in this volume.

[p] It should be remembered that the aim of this research was not to identify the dynamics of contagion as a phenomenon in itself, but to relate it to the territorial features of the areas considered. That is why the data have not been taken in their absolute values, but as a trend in epidemiological propagation. Data were interpolated with datasets related to different territorial features at different scales: of the province of Bergamo, Lombardy, Italy and Europe. Notwithstanding, serious shortcomings in the level of detail for publicly circulated data have been found. At a municipal scale, data across the various databases available online are the most inconsistent, since they were collected, processed and communicated independently and along dissimilar procedures by each Italian region.

[q] In addition to the highest absolute numbers of Covid-19 deaths, Lombardy recorded the highest variation rates and the highest mortality rates in the country. Based on mortality, the region is divided into three macro-areas: the provinces of Bergamo, Cremona, and Lodi with very high rates. Adjacent to these are the provinces of Brescia, Pavia, Mantua, Lecco, and Sondrio where the rate was higher than in other regions of Italy, but lower than the first three provinces. Finally, in the westernmost provinces (including the area of Milan) mortality impact was much lower and comparable to the Italian average. In this regard, see Chapter 2 of this volume.

conditions that may lead to a severe form of the disease.[r] These may be classified as: morpho-climatic aspects, which favor the onset of disease in relation to pollution; social features, such as urbanization and commuting, which affect the intensity of viral diffusion; and finally, a set of co-factors, which also contribute to the intensity and severity of infection, and are tied to the health and care system. While these factors are systematically addressed in the various chapters of this volume, it is worth recalling them here, in line with the theoretical approach of our research and to justify the three Italies model.

In the morphological and climatic configuration of the peninsula, the Po Valley appears as a depression between two mountain ranges (the Alps and the Apennines) with a concave shape that leads to the prevalence of a temperate, poorly ventilated continental climate (according to Köppen classification). This favors the stagnation of atmospheric currents and the concentration of precipitation in intermediate seasons, preventing the dispersal of polluting emissions and fine dust that could otherwise be swept away by winds or settled on the ground via precipitation.[s] The other Italian regions are also affected by geomorphological features, albeit in positive terms. In their case such features lead to favorable climatic regimes: proximity to the sea, for example, ensures mild temperatures and constant winds which facilitate pollutant dispersal (Mediterranean climate); the interior of the Italian Peninsula is affected by an Apennine climate, named after the homonymous orographic mountain range. Unlike the Northern Alps, the Apennines have limited elevation, which does not hinder air circulation but rather creates a favorable set of local conditions regarding temperature and rainfall.

As we turn to consider contagion in relation to the socio-territorial context, we cannot fail to see that the Po Valley is marked by high population density and a seamless urbanized *continuum*.[t] One need only note that the population of the Po basin exceeds 26 million inhabitants (corresponding to more than 40% of the Italian population), who are mainly distributed in the plains and valley bottoms, for an average density of about 450 inhabitants per square kilometer. We are faced with a populous conurbation that unites the main cities of Piedmont, Lombardy, Veneto, Friuli-Venezia Giulia, and Emilia-Romagna but is centered around the

[r]Research began with the question "Why Bergamo?" which is also the title of the first report we posted online in March 2020. This title was later taken up by Marco Cremaschi in a historical geography analysis: "Pourquoi Bergame? Le virus au bout du territoire" in June 2020.

[s]Of course, air pollution outcomes may not be quantified solely on the basis of emitters (factories, livestock, mobility, etc.), but depend also on how long pollutants remain in the air. In this regard, a study on the concentrations of NO_2 Tropospheric (nitrogen dioxide) on data extracted from the Sentinel-5P satellite were used to explain spatial variation in mortality cases for 66 administrative regions in four European countries. Sentinel-5P data pinpoints two hotspots in Europe, namely northern Italy and the Madrid metropolitan area, both areas with high Covid-19 contagion levels (Ogen, 2020). Furthermore, if we relate the climatic configuration described to pollution data and note that ¼ of the Italian population is concentrated in the Po Valley, the social impact of the health risk becomes evident.

[t]On the complexity of this settlement and the dynamics of Covid-19 diffusion within such contexts, see especially: Borruso et al. (2020); Djaiz (2020); Jackson (2020); Murgante et al. (2020); Lévy (1999, 2020); Lussault (2017, 2020a,b).

vast metropolitan area of Milan, which generates strong mobility and substantial flows of people and goods. This dynamic layout determines both a dense network of contacts in public spaces or hyper-places[u] and massive reticular movements via private individual mobility and collective public mobility, which greatly facilitate the viral spread by promoting multiple gatherings. The conspicuous presence of production enterprises in Lombardy and the substantial concentration of sector workers in the Milan area, in the northern metropolitan area, as well as in the areas of Bergamo and Brescia justifies the recurrence of region-based flows of people and the dense network of relational contacts between territories, which also extend to the neighboring areas of the Po Valley.[v]

The structural socio-territorial fragilities which favored contagion at the onset of the epidemic were compounded by major shortcomings of the health care system in the second phase of the epidemic, the phase of maximum morbidity, when a centralized model of care and the widespread shortage of regional facilities, at least for the Lombardy region, made it impossible to contain the infection. Even the social healthcare network aimed at the elderly and based primarily on Nursing and Residential Care Facilities (RSAs) proved easily vulnerable to the disease, both for the precarious health of elderly guests, and for the underlying management model, whereby health personnel was shared among facilities, thus engendering a rhizome-like spread of contagion.[w]

The third phase of contagion saw the adoption of lockdown measures imposed by Government Decree dated March 9, 2020 (*Further operational provisions of decree-law dated 23 February 2020, n. 6, containing urgent measures for the containment and management of Covid-19 epidemic emergency, to be enforced throughout the national territory*). While containment policies did lead to a progressive decline in infection rates, they fell short of eradicating the virus, partly because distancing regulations were ineffectively enforced only on parts of the production sector. This has had a negative impact both on the attempt to curtail commuting (which has relied exclusively on distance learning and smart working) and on the attempt to contain crowding on public transport and in hyper-places. The same predicament linked to housing density and commuting also occurred in other cities, outside the Po Valley area, such as Rome, Florence, Naples, or Genoa where contagion predictably increased, marking a break with extra-metropolitan areas.

The processing and analysis of data on the dynamics of Covid-19 contagion has pinpointed constants in viral diffusion, which enable us to advance possible lines of interpretation based on a space–time model. The first is that the morphological and climatic factors involved in pollution and the housing factors tied to the density and mobility of inhabitants did favor

---

[u]Hyper-places (*hyper-lieux*) are places characterized by marked density of both physical and virtual relations. They function simultaneously and intensively on multiple scales: they are inscribed in the local space, yet bound by global networks (Lussault, 2017).

[v]The exacerbation of contagion in Lombardy could be traced to the marked regional presence of hyper-connected urban systems and the high number of hyper-places. Weaknesses inherent in such configuration were highlighted by Covid-19, which affected precisely the fabric of relations and connections that unfold both on a local and global scale (Lussault, 2020a,b).

[w]The fragility of the healthcare system, characterized by serious shortcomings due to the high level of complexity and fragmentation in terms of skills and resources between institutional and non-institutional actors, is discussed in detail in Chapter 5 of the present volume.

contagion, especially in the first phase, when social distancing measures were yet to be adopted. The second, on the other hand, is based on the identification of weaknesses in the healthcare system of regions in Northern Italy, where Covid-19 propagation occurred both inside hospitals and inside Nursing and Residential Care Facilities. Regions in the north of Italy can, in fact, boast a wide range of healthcare facilities compared to the rest of the country, and these certainly played a role as the north struggled to bear the brunt of the epidemic.[x] However, Covid-19 has brought to the surface major shortcomings in the northern model for healthcare management and elderly care in RSAs. The fragility of the RSA model led to a rapid and severe spread of infection, in many cases by implosion: the metaphor of the lit match dropped in a haystack may serve to illustrate the wildfire-like propagation of the epidemic, impossible to contain.

While many aspects are yet to be investigated in full, a general lesson seems to emerge from the current experience, namely that complex societies (like ours) cannot possibly remain unprepared when faced with events of this magnitude. More importantly, as we struggle to defuse SARS-CoV-2, we also need to rethink our model of inhabiting places, with a view to eliminating or managing risk factors. This is in consideration of the warnings that come to us from many quarters, to the effect that similar threats may be expected to recur in our future.

1.2.4 Suggestions for a new territorial project

The analysis conducted so far has provided suggestions on how to address the questions posed by the current pandemic, such as the causes of its spread, the intensity of infection and, finally, the severity of its outcomes. Research was conducted from a point of view that is intentionally external to virology or epidemiology, in the awareness that many of the answers to these questions must be searched not solely by investigating the virus or its epidemiological features, but also by shedding light on the territorial factors that its distribution patterns entail. And researchers aimed not so much to dwell on exact contagion figures, but to highlight dissimilar developments or outcomes in relation to the regions involved, focusing on the ones which were most affected. That is why this study addresses mainly the northern part of the Peninsula, where the epidemic was more intense and most severe, and sets out to cross-reference geographical factors as probable co-determinants in the evolution of the Covid-19 epidemic.

Marked regional differences in the intensity and distribution of infection upheld our initial research hypothesis, namely that socio-territorial factors may have affected viral spread. A number of features of mobile and urbanized living that intervene in the spread of contagion were identified first, with a view to addressing aspects of morphology and climate and their effects on pollution. Subsequently, specific conditions were sought within this initial set of

[x]The dynamics of Covid-19 diffusion recorded in autumn 2020 differ from those observed during the first wave. While in the months of March and April 2020 contagion was by and large limited to northern provinces, which were very severely affected, in the months of October and November 2020 it reached provinces in Central and Southern Italy. These feature large metropolitan clusters, which have recorded intense viral outbreaks also linked to shortcomings in the staying power of the regional healthcare system. The observations rely on data provided by the Ministry of Health and the Italian Department of Civil Protection.

data that could possibly lead to overcrowding, which in turn would make isolation difficult and eventually expose shortcomings in containment responses. These combined factors, which we defined as "territorial fragilities," will have to be acted on in order to design a new territory suitably fitted to deal with epidemics.

Reflexive cartography played a key role in this context, precisely because—by cross-referencing multiple indicators within the same datasets—it laid the foundations for researching possible determinants for the dissimilar patterns of viral spread and highlighted different modes of aggression in viral propagation. The social sense of territory was presented via digital mapping, which makes it possible to grasp and effectively convey the multi-layered significance of the territory and the complex development of the epidemic, thereby bringing out the so-called "fragilities of contemporary living" (Casti, 2020, p. 75). These refer to structural aspects that favored the epidemic spread and include, for instance, intense daily flows of people, population density, vulnerability of the welfare and health-care system as well as atmospheric pollution. These fragilities may be considered the basis from which to rethink territorial policies both during and after the Covid-19 health emergency.

To draw generalizations from this study may be ill-advised. Nonetheless, the notions that underlie the present approach—namely that current patterns of living in a mobile and urbanized world emerge in the intertwining of nodes and connections, which are shaped by the dynamic interaction of inhabitants and inevitably favor the spread of epidemics—could reasonably be extended beyond the Italian context, to a large part of the world. To have outlined risks and vulnerabilities in this form of living is nonetheless an act of awareness, for these will undeniably have to be addressed and examined with care, if we intend to (re)think territory in the light of the pandemic event that keeps raging, worldwide, to this day.

References

Borruso, G., Balletto, G., Murgante, B., Castiglia, P., Dettori, M., 2020. Covid-19. Diffusione spaziale e aspetti ambientali del caso italiano. Semestrale di Studi e Ricerche di Geografia 32 (2), 39–56.

Casti, E., 2015. Reflexive Cartography. A New Perspective on Mapping. Elsevier, Amsterdam.

Casti, E., 2020. Geografia a 'vele spiegate': analisi territoriale e mapping riflessivo sul Covid-19 in Italia. Documenti Geografici 1, 61–83.

Consolandi, E., Rodeschini, M., 2020. La cartografia come operatore simbolico: il contagio del Covid-19 in Lombardia. Documenti Geografici 1, 711–724.

Djaiz, D., 2020. La mondialisation malade des ses crisis? Le Grand Continent. accessed October 2020 from https://legrandcontinent.eu/fr/2020/03/23/coronavirus-mondialisation-david-djaiz/.

Jackson, M.O., 2020. Comment se diffuse un virus? Le Grand Continent. accessed October 2020 from https://legrandcontinent.eu/fr/observatoire-coronavirus/.

Lévy, J., 1999. Le tournant géographique. Penser l'espace pour lire le monde. Editions Belin, Paris.

Lévy, J., 2020. L'humanité habite le Covid-19. AOC. Analyse, Opinion, Critique, accessed October 2020 from https://aoc.media/analyse/2020/03/25/lhumanite-habite-le-covid-19/.

Lussault, M., 2017. Hyper-lieux. Les nouvelles géographies de la mondialisation. Seuil, Paris.

Lussault, M., 2020a. Le Monde du virus – retourn sur l'eproure de confinement. AOC. Analyse, Opinion, Critique. accessed October 2020 from https://aoc.media/analyse/2020/05/10/le-monde-du-virus-retour-sur-lepreuve-du-confinement/.

Lussault, M., 2020b. Chroniques de géo' virale. Ecole urbaine de Lyon/Editions deux-cent-cinq, Lione.

Murgante, B., Borruso, G., Balletto, G., Castiglia, P., Dettori, M., 2020. Why Italy first? Health, geographical and planning aspects of the Covid-19 outbreak. Sustainability 12 (5064). https://doi.org/10.3390/su12125064.

Ogen, Y., 2020. Assessing nitrogen dioxide (NO_2) levels as a contributing factor to the coronavirus (Covid-19) fatality rate. Sci. Total Environ. 726. accessed October 2020 from https://www.ncbi.nlm.nih.gov/pmc/articles/PMC7151460/pdf/main.pdf.

Zhou, C., et al., 2020. Covid-19: challenges to GIS with big data. Geogr. Sustain. 1 (1), 77–87.

Further reading

Castaldini, D., Marchetti, M., Norini, G., Vandelli, V., Zuluaga Vélez, M.C., 2019. Geomorphology of the Central Po plain, northern Italy. J. Maps 15 (2), 780–787.

Casti, E., 2000. Reality as Representation. The Semiotics of Cartography and the Generation of Meaning. Bergamo University Press, Sestante, Bergamo.

Connolly, C., Ali, S.H., Keil, R., 2020. On the relationships between Covid-19 and extended urbanization. Dialogues Hum. Geogr. 10 (2), 213–216.

Cremaschi, M., 2020. Pourquoi Bergame? Le virus au bout du territoire. Métropolitiques. accessed October 2020 from https://www.metropolitiques.eu/Pourquoi-Bergame-Le-virus-au-bout-du-territoire.html.

Franch-Pardo, I., Napoletano, B.M., Rosete-Verges, F., Billa, L., 2020. Spatial analysis and GIS in the study of Covid-19. A review. Sci. Total Environ. 739, 1–10.

Fraser Taylor, D.R., 2006. The theory and practice of cybercartography: An introduction. In: Fraser Taylor, D.R., Lauriault, T. (Eds.), Cybercartography, Theory and Practice. Elsevier, Amsterdam, pp. 1–13.

Fraser Taylor, D.R., 2019. Cybercartography revisited. In: Fraser Taylor, D.R., Anonby, E., Murasugi, K. (Eds.), Further Developments in the Theory and Practice of Cybercartography. Elsevier, Amsterdam, pp. 3–23.

Lussault, M., 2007. L'Homme spatial. La construction sociale de l'espace humain. Seuil, Paris.

Mooney, P., Juhàsz, L., 2020. Mapping Covid-19: How web-based maps contribute to the infodemic. Dialogues Hum. Geogr. 10 (2), 265–270. https://doi.org/10.1177/2043820620934926.

Söderström, O., 2020. Smart city citoyenne et pandémie. AOC. Analyse, Opinion, Critique. accessed October 2020 from https://aoc.media/analyse/2020/06/01/smart-city-citoyenne-et-pandemie/.

1.3

Evolution and intensity of infection in Lombardy

Elisa Consolandi

1.3.1 Premise

As is now sadly known, Lombardy is the Italian territory most affected by the SARS-CoV-2 virus. The first outbreaks of the epidemic in Italy were identified in this region—specifically, in the province of Lodi and, later, in that of Bergamo—and it is precisely these areas that recorded the highest number of infected people during the first epidemic wave (February–June 2020). While a cursory estimate may classify the whole of Lombardy as a high-contagion region with an infection rate far above the national average, a detailed assessment in fact brings up areal differences that must be taken into account in order to investigate any possible correlations with socio-territorial factors. Because of this, contagion data were geolocated and their evolution traced by cross-reference to population distribution data. We may follow the space–time evolution of contagion by interpreting our reflexive map.

1.3.2 The Lombard territory: Distribution and temporal evolution of infection

As we turn to examine the spread of Covid-19 contagion in Lombardy,[a] we should start by noting that within barely a few weeks from the epidemic onset (indicatively, from the

---

[a] It should be noted that analysis of the pandemic event in Italy has faced difficulties, to do with the retrieval and, above all, the reliability of data about infections at different scales, from the local to the national one. For the purpose of cartographic production, data made available by institutional sources, such as the Ministry of Health and the Higher Health Institute (*Istituto Superiore di Sanità*) were used, in addition to data provided by individual regions and other institutions (such as, for instance, the health protection agencies and regional health and welfare bodies). Despite this, serious lack of information on a municipal scale remains.

Topographic map
(a)

Anamorphic map relative to the number of residents
(b)

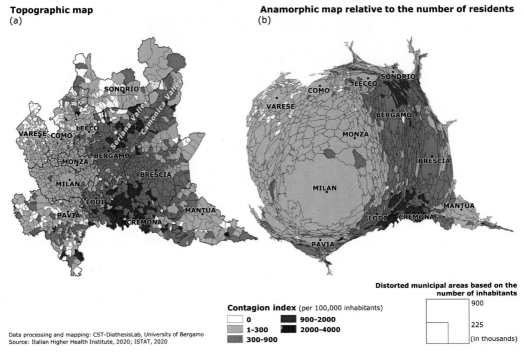

Contagion index (per 100,000 inhabitants)

☐	0	■	900-2000
☐	1-300	■	2000-4000
☐	300-900		

Distorted municipal areas based on the number of inhabitants

900

225

(in thousands)

Data processing and mapping: CST-DiathesisLab, University of Bergamo
Source: Italian Higher Health Institute, 2020; ISTAT, 2020

FIG. 1.6 Lombardy: distribution of the Covid-19 infection in relation to the resident population as of 23 March 2020.

beginning of March to the end of April 2020[b] the situation in the region worsened rapidly, especially in its central-eastern section. As illustrated in Fig. 1.6, discontinuous distribution of the contagion rate emerged from the beginning and was addressed by using specific cartographic techniques. In fact, the map based on initial infection data—as of 23 March 2020—outlines the region of Lombardy both (a) on the basis of standard topographical measurement; and (b) in an anamorphically distorted view relative to the number of inhabitants.[c] Both outlines share the same index, sorted within a *range,* and color-coded accordingly. In the first image (a) the index was spread over the administrative extension of the municipalities; in the second (b), instead, the index was made to match the expansion or contraction of each

[b]It should be noted that cartography data processing is the result of constant observation on the part of Territorial Studies Center (*Centro Studi sul Territorio,* CST) research group at the University of Bergamo. Since March 2020, CST researchers have been investigating the geographical significance of the Covid-19 infection, on the basis of a reflexive mapping model that brings into renewed focus the spatial dimension of geo-analysis (Casti, 2020; Consolandi and Rodeschini, 2020).

[c]Distortion was obtained via anamorphosis, and relies on the processing of shapefiles through ScapeToad, a cartogram application that enables researchers to adapt map surfaces to user-defined variables without altering their topological relations. Municipalities on the map were distorted in proportion to the absolute number of resident inhabitants. The program was created by the Chôros Laboratory of the École Polytechnique Fédérale in Lausanne (Switzerland). The software is open-source and is available at the following link: http://scapetoad.choros.place/ (last accessed: December 2020).

area based on the high or low number of inhabitants. The resulting communicative outcomes differ markedly: in the first, color-coding is the only indicator of dissimilar infection distributions; in the second, it clearly emerges that the most affected area is the central-eastern part of Lombardy, which is also the least inhabited area when compared to the metropolitan area of Milan.[d]

Before analyzing results, it may be useful, however, to dwell briefly on this double representation technique, to underline its different cartographic and communicative potentials. It is well known that, by faithfully representing the surface of the territory, topographic metrics[e] enables researchers to localize a phenomenon. However, topographic metrics does neglect the social significance of territory because it does not relate territory to the people who inhabit it. Topographical representation may be said to *present* spatialized data, while it stops short of *grasping* its meaning, since "the possibility of juxtaposing one's own perceptual experience of the object represented is ruled out and what is envisaged is accepted uncritically" (Casti, 2015, p. 82). Anamorphic cartography,[f] instead, opts to neglect criteria of metric accuracy in order to enhance the social aspects of the phenomenon it represents—in this case, the epidemic spread. To do this, data relating to the phenomenon is combined with other information, always from the social sphere.[g] Using distortion techniques, the surface of the territory—of the map base—is made to correspond to the number of individuals who inhabit it, thereby highlighting the magnitude of the social risk of being infected. This communicative outcome is possible because the communicative level changes: from a surface level, based on reference, we shift to a connotative level, centered on the transmission of social values. A connotational model underlines the social importance of contagion instead of simply localizing it (Casti, 2015). In short, manipulation of the areal surface corresponding to each municipality and the use of color-coding yield information on the different levels of infection severity in Lombardy. And by thus tracing the evolution of contagion we reflect on the dynamics that may have led to its increase.

Returning now to the contents of the anamorphic map, we should reiterate that the Mideast ridge presents contagion peaks in the province of Lodi and in Bergamo, where, as mentioned earlier, the outbreaks of Codogno and Nembro-Alzano Lombardo occurred.[h] On the contrary,

[d]Except for the province of Mantua, which initially seemed to have been only slightly affected.

[e]Topographic metrics is a representational system based on Cartesian principles and on the representation of Euclidean space, which are "brought together into a measuring system for distance that is not concerned with the representation of the quality of objects but with standardizing it, preserving their relationship, their size" (Casti, 2015, p. 72).

[f]Anamorphosis is a technique devised in the Renaissance period with the aim of distorting figures; the image to be represented is deliberately distorted in order to "emphasize the inadequacy of the image to render the meaning of things and, consequently, to warn not to rely on appearance unthinkingly" (Casti, 2015, p. 249).

[g]Cartographic representation arguably plays a vital role in communications on the Covid-19 pandemic, since it works effectively as a symbolic operator (Casti, 2015, pp. 17–20), actively intervening in the production of knowledge and the implementation of actions aimed at epidemic containment.

[h]In general, initial outbreaks in Italy were recorded in Veneto and Lombardy. However, while outbreaks in Veneto (namely in the town of Vo' Euganeo in the Padua area) were quickly circumscribed, contagion in the Codogno areas (in the province of Lodi) and—especially—in the area of Alzano Lombardo-Nembro (in Bergamo) propagated and extended rapidly both via proximity and via reticularity. An in-depth discussion of this development may be found in Chapter 5 of this volume.

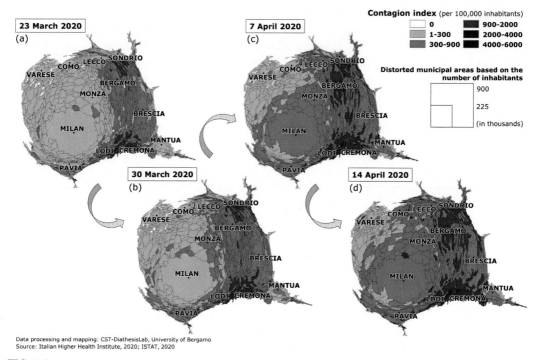

Data processing and mapping: CST-DiathesisLab, University of Bergamo
Source: Italian Higher Health Institute, 2020; ISTAT, 2020

FIG. 1.7   Lombardy: evolution of the Covid-19 infection in relation to resident population from 23 March to 14 April 2020.

the vast metropolitan area of Milan, with Monza and the territories of Como and Varese, seems affected by an average contagion rate, although in the south—along the banks of the Po river—contagion is similar to that of the eastern area, that is, of medium proportions.

Distribution is confirmed and may be found to worsen if we trace the evolution of contagion from March to April[i] in four anamorphic sequences (Fig. 1.7): in the first (a), the Lombard towns recording more infections are located to the east of the median regional ridge which—in addition to the municipalities already mentioned in the provinces of Lodi and Bergamo—includes Cremona and some municipalities in the Brescia area; in the second map (b) contagion intensity may be seen to have extended to the entire eastern part of the region, where almost all municipalities record an incidence that exceeds 300 cases per 100,000 inhabitants. In addition to this, there is an increase in contagion also in western municipalities, albeit more contained. It is in the third map (c), however, which reports contagion

[i]Data provided by the Higher Health Institute (https://www.epicentro.iss.it/coronavirus/sars-cov-2-sorveglianza-dati) relate to infected people by municipality and, based on the evolution of the infection, they were updated until April 14, 2020. After that date, data on infection on a municipal scale for the Lombard territory were no longer provided. Municipalities marked in white represent a number of infected persons equal to zero. However, this data also includes municipalities for which the number of infected persons is unspecified. Also, our map indicates that some municipalities around the metropolitan area differ from the prevailing rate of the area they belong to. Reasons for this variation are yet to be investigated.

data after 1 week, that widespread intensification of contagion is clearly observed. This also affects the metropolitan area of Milan with the northern areas of Varese and Como, even though contagion does not reach the maximum levels recorded in municipalities to the east. In the last representation (d), the whole region appears severely infected, except for the Varese area.[j] As of April 14, a growing progression of the epidemic affecting the Como area and the western municipalities of the region may be observed, while the further intensification is witnessed in the Lodi-Cremona, the Bergamo, and the Mantua areas.

Initial assessment of discontinuities in the distribution of infection, which seems to progress over time while also maintaining different degrees of intensity, leads to the conclusion that highest Covid-19 spread rates may actually be expected in peripheral areas as opposed to large urban centers. Although periurban territory partakes of the region-wide conurbation it does not strictly fall within the metropolitan area of Milan. Similarly, a higher incidence of infection did not initially affect intermediate urban centers such as Bergamo or Brescia, but disrupted municipalities in their respective periurban areas,[k] which consist generally of settlements of medium or small size. Such data seem to confute the hypothesis whereby population density is a major determinant of the Covid-19 epidemic, possibly suggesting instead that although the kind of overcrowding density entails undoubtedly facilitates viral spread, other factors need to be traced to make sense of infection when its proportions are elevated.[l] The areas of Lombardy that have endured severe contagion are mainly periurban and present average population densities. However, they are characterized by very vibrant

[j]It should be noted that the Varese area, which like the area of Milan seemed to have been largely spared or, at least, only tangentially affected by the contagion in the first wave, became a hotbed of infection in the second wave, which began in October 2020. Contagion in the province of Varese was around 1800 infected people (with an index of 320 sick persons per 100,000 inhabitants), while in October this figure increased considerably, to a count of nearly 4700 infected people (i.e., 720 infected persons per 100,000 inhabitants). Virologists tend to interpret such surge as a result of the fact that those who have had Covid-19 may not be thoroughly immunized, as possibly indicated by the recent data published online (Tillett et al., 2020).

[k]Indeterminacy persists as to the exact meaning of "periurban," and definitions vary. For some, periurbanization is the process whereby new urban settlements are established more or less close to large centers or major communication routes, in accordance with a diffuse morphology model, which "tends to turn itself into a city" (Dézert et al., 1991, p. 25). Others, such as Lévy and Lussault (2014) claim that "periurbanization [...] is but one of the spatial expressions of diffusion, which is associated here with the development of areas set off from a pre-existing agglomeration" (p. 2).

[l]On the issue of whether population density is a factor that facilitates the spread of infection a range of opinions exists. Desai (2020), among others, claims that as a result of an increase in urban density the constant growth in the number of people inhabiting a given area puts a strain on public services and thus also on the healthcare system at a time of crisis, such as the current epidemic. In May 2020, a study by INSEE—the Institut national de la statistique et des études économiques—(Gascard et al., 2020) established a link between municipal density and the Covid-19 death rate recorded throughout France. Ourfeuil (2020a,b) agrees instead with other researchers who argue that population density is not the cause of viral spread, because diffusion may be facilitated instead by other factors, such as: urban poverty, congestion of public transport, connections, and contact intensity. In fact, high relational densities may lead to greater viral spread and cause very high mortality rates in large metropolitan areas. As early as the first months of 2020, however, Jacques Lévy voiced an alternative opinion with regard to Covid-19 when he noted that periurban spaces seem in fact to emerge as the first centers to be affected by the pandemic all over the world, at least at the time the virus reached the first centers in March (Lévy, 2020).

networks of contacts and exchanges. These areas rely heavily on internal commuting, a practice that affects contacts both via urban proximity and via reticularity, due to intense movement.[m]

Conversely, it should be noted that, at least as regards Italy, in the subsequent, acute phase of Covid-19 infection (corresponding roughly to the month of April 2020), contagion intensified especially in urban areas. So, for instance the metropolitan area of Milan[n] was hard hit only later than other areas in eastern Lombardy.

In our research, diffusion of Covid-19 in Lombardy was therefore investigated in relation to the distribution, evolution, and composition of the population, which allowed us to derive more detailed information on epidemic propagation. At the same time, cartographic processing made it possible to compare different datasets on the infection, namely the absolute number of viral cases and the contagion index recorded for each Lombard municipality. These in turn ensured a more articulate understanding of the quantitative extent of the epidemic phenomenon. Graphically, *extrusion* was used to assign dimensional relief to municipal surfaces and obtain three-dimensional shapes which provide an immediate visualization of the absolute number of infected persons recorded at specific intervals in each area of Lombardy. The contagion index was subsequently superimposed using color-coded gradients which convey information on contagion risk. Since the latter factor is calculated in relation to the resident population for each municipality, it effectively conveys the social impact of contagion side by side with a purely quantitative assessment of its reach.

Data visualization in Fig. 1.8 indicates that areas most affected by Covid-19 in absolute terms do not coincide with those which recorded highest contagion rates. These latter areas coincide mainly with infection hotbeds, marked on the map with a deflagration symbol and, as we have seen, these do not correspond to large urban areas of Lombardy. The map shows that the greatest number of infections in absolute terms occurred in the cities of Milan, Bergamo, and Brescia, in which, however, viral incidence was considerable but not as extreme as in the city of Cremona, Lodi and in some small southern municipalities. Even though these small municipal areas have little or no *extrusion* and graphically suggest that the absolute number of infected people is relatively low, they also convey the information that the risk of catching the disease for their respective areas is the highest in Lombardy. Self-referential information[o] drawn from this map confirms our initial assessment to the effect that there seems to exist a mismatch between population density and contagion intensity, which in this case is highlighted by the fact that the provinces are located in the periurban area around metropolitan Milan, precisely where the first Lombard outbreak was recorded. Albeit in very tentative and provisional terms, evidence suggests that low population density is not

---

[m] On this issue, see Chapter 3 of this volume.

[n] As of April 7, 2020, the metropolitan area of Milan recorded a contagion rate of between 300 and 900 patients per 100,000 inhabitants, while Bergamo, Cremona and Lodi had a higher contagion rate, corresponding to 900 infected persons per 100,000 residents.

[o] In this context, maps become "complex communicative systems, which internally develop sets of self-referential information. These in turn give substance to maps representational power" (Casti, 2015, p. 17). Maps are then taken as symbolic operators, able to intervene actively in the production of knowledge over a specific phenomenon, the social meaning of which they are able to render.

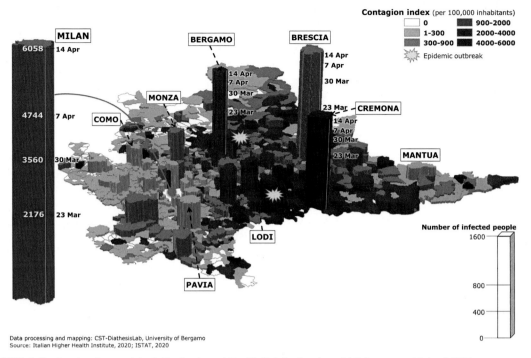

FIG. 1.8   Lombardy: municipal distribution of Covid-19 infection from 24 February to 14 April 2020.

a predictor of infection intensity, nor does it preclude initial onset. Thus, when cross-referenced to absolute contagion figures, data concerning the viral incidence of Covid-19 in Lombardy yield a different picture to the one derived solely via geolocalization as to the number of sick persons in a given area. Reflexive cartography trumps geolocalized representation.

As for the deadliest outcome of Covid-19 contagion in Lombardy, the high mortality rate recorded among the elderly cannot possibly have been due solely to their status as high-risk subjects, as members of the population who are exceptionally weak and therefore vulnerable to disease. That is because Lombardy presents the same age distribution as other Italian regions, which all reflect population aging affecting the entire country.[P] Also, when the data are broken down according to provinces, the lethal outcomes are seen to vary considerably.

[P]The metropolitan city of Milan is the most densely populated area, followed by the provinces of Bergamo and Brescia. Lodi and Sondrio, on the other hand, count the lowest number of inhabitants. As regards its setup, the population of Lombardy presents the same profile as other Italian regions: the widest age bracket includes residents between 40 and 59 years; while lower age groups mark a sharp decline—indicative of a low birth rate—and higher age groups follow a progressive, natural trend which attests to high life expectancy and indicates aging (see the Italian Office for National Statistics, ISTAT website: http://demo.istat.it/).

Mortality rates recorded in the areas of Bergamo, Cremona, and Lodi are so high that it would simply be unreasonable to ascribe them solely to the presence of a large elderly population.[q] These outcomes must be traced to other socio-territorial determinants, such as—for instance—the lack of adequate health care, as research seems ever more strongly to suggest.[r]

1.3.3 Concluding remarks

The research we carried out in this section to investigate the causes of the intensity and severity of Covid-19 infection involves attention to the tight connections between places, which determined viral propagation both by proximity and by reticularity. The case of Lombardy, characterized by mobility and intense urbanization, validates this working hypothesis, and demonstrates that Covid-19 infection surged in tightly connected periurban areas, where propagation occurs both (a) by reticularity, since such areas rely on intense commuting and on a dynamic network of exchanges; and (b) by proximity to outbreaks, which were not promptly isolated.

Via the use of reflexive maps, a set of social factors that may have affected the epidemic phenomenon was brought to the fore, with a view to grasping the dynamics of diffusion and interpreting discontinuities in the distribution of contagion rates. Once again, cartography proves an invaluable communicative tool for coming to terms with the intricate dynamics of Covid-19 propagation. To employ alternative and innovative graphic formats, for instance anamorphosis or tridimensionality, to the field of geo-representation, is far more than simply proposing more captivating or appealing models of visualization. Reflexive maps are effective tools of communicative mediation, capable of producing self-referential data which contribute to a more articulate understanding of the phenomena maps represent and suggest valuable interpretative keys for their management.

Along these lines, it should be clear that the distribution and regional evolution of Covid-19 contagion in Lombardy in relation to its population; the assumption of Lombardy as a prototype of mobile and urbanized living; and the cross-referencing to municipal data about the resident population have led our research to produce a multi-scale framework. And such framework forcefully reiterates that these territorial features played a crucial role in viral propagation.

---

[q]The impact of mortality is higher than in other regions of Italy even in the province territories of Brescia, Pavia, Mantua, Lecco and Sondrio. However, in westernmost provinces (including the metropolitan area of Milan) death rates are lower and comparable to the Italian average. The lethal outcomes recorded for the months of March and April 2020 in the aforementioned provinces are up to four times higher than the death rates recorded in previous years for the same period. See Chapter 2 of this volume.

[r]The swift spread of contagion and the severity of the pandemic in Lombardy were cross-referenced to features of the regional health care system and to the number of elderly people housed in Nursing and Residential Care Facilities (RSAs) located in the region. Analysis exposed the fragility of the healthcare system, characterized by serious shortcomings attributable to the high level of complexity and fragmentation, in terms of skills and resources, between institutional and non-institutional agencies. The issue is discussed at length in Chapter 5 of this volume.

To be sure, Lombardy is also a region steeped in the operational dynamics of globalization. It hosts a high number of facilities and services and relies on spatial setups which play prominent relational roles such as hyper-places (airports, stations, or shopping centers, in short, dynamic public spaces), which are determined by multiple connections, both actual and virtual. That all these features should render such spatial configurations vulnerable has dramatically been brought to our attention by the unexpected onset of the Covid-19 epidemic, and by the forced limitations which worldwide contagion has placed onto this intricate tangle of global connections. The economic crisis that ensues underscores the fragility of contemporary living.

References

Casti, E., 2015. Reflexive Cartography. A New Perspective on Mapping. Elsevier, Amsterdam.

Casti, E., 2020. Geografia a 'vele spiegate': analisi territoriale e mapping riflessivo sul Covid-19 in Italia. Documenti Geografici 1, 61–83.

Desai, D., 2020. Urban densities and the Covid-19 pandemic: Upending the sustainability myth of global megacities. In: ORF Occasional Paper, 244. Observer Research Foundation, New Delhi.

Consolandi, E., Rodeschini, M., 2020. La cartografia come operatore simbolico: il contagio del Covid-19 in Lombardia. Documenti Geografici 1, 711–724.

Dézert, B., Metton, A., Steingerg, J., 1991. The périurbanisation en France. Sedes, Paris.

Gascard, N., Kauffmann, B., Labosse, A., 2020. 26% de décès supplémentaires entre début mars et mi-avril 2020 : les communes denses sont les plus touchées. In: INSEE Focus, 191. accessed December 2020 from https://www.insee.fr/fr/statistiques/4488433#consulter.

Lévy, J., 2020. L'humanité habite le Covid-19. AOC. Analyse, Opinion, Critique. accessed October 2020 from https://aoc.media/analyse/2020/03/25/lhumanite-habite-le-covid-19/.

Lévy, J., Lussault, M., 2014. Périphérisation de l'urbain. Espaces Temps. accessed October 2020 from www.espacestemps.net/articles/peripherisation-de-lurbain.

Orfeuil, J.P., 2020a. Épidémie de Covid et territoires. In: Fonciers en débat. accessed December 2020 from https://fonciers-en-debat.com/epidemie-de-covid-19-et-territoires.

Orfeuil, J.P., 2020b. Densité et mortalité du Covid-19 : la recherche urbaine ne doit pas être dans le déni! In: Métropolitiques. Accessed December 2020 from https://metropolitiques.eu/Densite-et-mortalite-du-Covid-19-la-recherche-urbaine-ne-doit-pas-etre-dans-le.html.

Tillett, R.L., et al., 2020. Genomic evidence for reinfection with SARS-CoV-2: a case study. Lancet Infect. Dis. accessed November 2020 from https://doi.org/10.1016/S1473-3099(20)30764-7.

Further reading

Fraser, T.D.R., 2019. Cybercartography revisited. In: Fraser, T.D.R., Anonby, E., Murasugi, K. (Eds.), Further Developments in the Theory and Practice of Cybercartography. Elsevier, Amsterdam, pp. 3–23.

Lussault, M., 2020a. Le Monde du virus—retourn sur l'eprouve de confinement. AOC. Analyse, Opinion, Critique. accessed October 2020 from https://aoc.media/analyse/2020/05/10/le-monde-du-virus-retour-sur-lepreuve-du-confinement/.

Lussault, M., 2020b. Chroniques de géo' virale. Ecole urbaine de Lyon/Editions deux-cent-cinq, Lione.

Mooney, P., Juhàsz, L., 2020. Mapping Covid-19: how web-based maps contribute to the infodemic. Dialogues Hum. Geogr. 10 (2), 265–270. https://doi.org/10.1177/2043820620934926.

Sun, Z., Zhang, H., Yang, Y., Wan, H., Wang, Y., 2020, 1–7. Impacts of geographic factors and population density on the Covid-19 spreading under the lockdown policies in China. Sci. Total Environ. 746.

CHAPTER

1.4

Contagion and local fragilities in Bergamo and the Seriana Valley

Elisa Consolandi and Marta Rodeschini

1.4.1 Premise

Besides being the Italian region most intensely affected by contagion, the mid-east section of Lombardy also recorded the onset of the first Covid-19 outbreaks,[a] which were localized in the province of Lodi and around Bergamo (namely, in the Seriana Valley). This prompted us to zero in on the Seriana Valley with the aim of illustrating the features that the epidemic has taken up in the province of Bergamo. Data relating to the SARS-CoV-2 virus were cross-referenced to a number of socio-territorial aspects[b] to test the hypothesis that the housing and functional complexity of the territory under consideration did affect the timing and severity of infection.

The method our analysis relies on is strongly linked to our research perspective,[c] which considers contemporary living as based and identified by mobile and urbanized patterns (Casti, 2020). On this assumption, we propose a set of cartographic models in the role of symbolic operators, aimed simultaneously at (1) representing a specific phenomenon and

---

[a] The epidemiological term "outbreak" indicates the location from which the pathogen spreads, starting from a group of individuals who have been subjected to shared virological exposure. For an in-depth discussion of the initial phase of viral spread, see Chapter 5 of this volume.

[b] Albeit rather fragmentary and discontinuous (Casti, 2020), the infection data we processed come from institutional sources, namely from the Italian Department of Civil Protection—Ministry of Health (http://opendatadpc.maps.arcgis.com/apps/opsdashboard/index.html#/b0c68bce2cce478eaac82fe38d4138b1), from the Lombardy Region (https://experience.arcgis.com/experience/0a5dfcc103d0468bbb6b14e713ec1e30/) and Higher Health Institute (https://www.epicentro.iss.it/coronavirus/sars-cov-2-sorveglianza-dati).

[c] On our research approach, see the introduction to this volume by Emanuela Casti.

(2) recovering its social significance via a perspective centered on reflexivity (Casti, 2000, 2015). The final goal is to shed light on some of the factors that facilitated the spread of the Covid-19 epidemic.

1.4.2 The province of Bergamo as a case study

The province of Bergamo extends between the provinces of Milan and Brescia in the center-east part of the Lombardy region. Its territorial boundaries are determined by the Orobie alpine mountain range and by its hydrographic basins, namely the Serio, Brembo, Adda and Oglio rivers. In addition to being important waterways, these rivers outline provincial demarcations. The Bergamo province has a population density of 405.2 inhabitants per square kilometer and over one million residents, distributed in the 243 municipalities it comprises.[d] Bergamo is the fourth province of Lombardy by extension.

The Bergamo area is part of a geographical layout marked by wide morphological and settlement diversity. Its territory lies at the center of a highly urbanized macro-region and inside an ecosystem that ranges from alpine landscapes, with peaks up to 3000 m high to agricultural plains; from hills to lakeside environments. Three specific contexts, relatively consistent in terms of their geographical, social and economic features, may be identified within this frame: the northern belt of valleys, the central and hilly belt which hosts the capital city of Bergamo, and finally the lower plains.[e] The Bergamo area lies at the center of one of the most developed geographical territories in Europe: proximity to Milan and close connections to high-density industrial and manufacturing areas were decisive conditions for Bergamo's economic development. The city of Bergamo[f] is located at the epicenter, in the shape of a

---

[d]ISTAT data as of January 1, 2020 (http://demo.istat.it/index_e.php).

[e]The northern belt occupies 64% of the entire mountain province surface: the Brembana Valley (crossed by the Brembo river), the Seriana Valley (Serio) and the Cavallina Valley (Cherio) and other minor valleys, such as the Imagna Valley, the Scalve Valley and the Serina Valley. The territory also hosts key tourist centers for Bergamo (among others, Monte Isola, Trescore Balneario, Clusone and San Pellegrino Terme). The central hilly belt, which extends for 70 km from the Adda river to Lake Iseo, is traversed by the A4 motorway. Even though its surface area is smaller, geographically this belt encloses the Bergamo area's nerve centers. The central area hosts the city of Bergamo, but also the town of Dalmine and the Tenaris steel mill, Stezzano and the Kilometro Rosso business incubator, Brembo SpA, Italcementi SpA and the Mario Negri Institute. To these attractors we should add the San Martino Valley, the Colli di Bergamo park, the Valcalepio area and, finally, the Orio al Serio Caravaggio Airport, which has of late played a leading role in the development of the entire province. The low plain is the southernmost part of the Bergamo territory and includes the resurgence belt. The most prominent cities are, among others, Treviglio, Romano di Lombardia and Caravaggio. This is an area characterized by the presence of medieval villages that have preserved their historical tradition over time and have strong agriculture-based features, favored by a dense network of irrigation ditches and canals fed by the waters of the Adda, Oglio and Serio rivers. The area of the plains (la Bassa) also hosts Crespi d'Adda, one of the oldest and most valuable examples of a workers' village in Europe and a UNESCO World Heritage Site since 1995 (De Angelis et al., 2019, p. 9–10; Ferlinghetti, 2015, pp. 111–140).

[f]In particular, the city of Bergamo was originally set up along the south-eastern tip of a hilly ridge, while the cluster of the historic center of Città Alta (Upper Town) developed along an undulating ridge, most of which has been progressively leveled by anthropic action (Ferlinghetti, 2015, p. 112).

well-established cluster that comprises a compact and densely inhabited community, seamlessly joined to a thick conurbation made up of all the neighboring municipalities within a settlement pattern in which the lower Seriana Valley plays a role of primary importance. Therefore, within the vast urbanized area that distinguishes the Po Valley, the city of Bergamo is characterized by intense exchanges and high inhabitant mobility both locally and on a national and international scale.[g] The Bergamo area also features a leading airport *hub* for the transport of people and industrial or consumer goods. It is a globalized and vibrant settlement that has built up an intricate set of interactions and socio-cultural connections[h] over time.

With regard to the Covid-19 epidemic, Bergamo ranks as one of the Italian provinces where the largest number of cases has been recorded.[i] The map in Fig. 1.9 was processed by interpolating the data on the number of infected persons[j] with data relating to the resident population divided by municipalities. The technique used to convey information is anamorphosis, which uses surface distortion based on the number of inhabitants in each Bergamo municipality to effectively outline the massive spread of contagion. Municipalities are not shown here on the basis of their geographical extension. Rather, they are visualized in relation to their resident population, so that one can immediately identify both the size of inhabited clusters and the high population density across the plains. Conversely, areas

---

[g]Work and business-related exchanges, tied to the crucial industrial sector located in this area, are key elements for quantifying people flows. For further information see Chapter 3 of this volume.

[h]Over the past few years, the international airport "Il Caravaggio" at Bergamo-Orio al Serio has witnessed an exponential growth in air traffic, driven largely by low-cost mobility. In 2019 it hosted 13,857,257 passengers, ranking as one of the top airport hubs nationwide (Assoaeroporti data, http://www.assaeroporti.com/, accessed on 24 February 2021). To this must be added intense commuting between the various urban centers, such as Bergamo, Brescia and Milan, which—in addition to major provincial agencies—also host leading companies, universities, and research institutes of national and international relevance. Tourism has also increased in recent years, and Bergamo is no longer merely a stopover site but a veritable tourist destination.

[i]By March 8th, when the first Prime Ministerial Decree under the title "Misure urgenti in materia di contenimento e gestione dell'emergenza epidemiologica da Covid-19" (*Urgent measures regarding the containment and management of Covid-19 epidemic emergency*) finally came into effect, forcing *lockdown* in Lombardy and in other 14 provinces of Northern Italy, the province of Bergamo already counted 997 Covid-19 cases. This number rose sharply within 2 days: on 10 March—when a second decree was issued with the title "Ulteriori misure in materia di contenimento e gestione dell'emergenza epidemiologica da Covid-19 sull'intero territorio nazionale" (*Further measures regarding the containment and management of the Covid-19 epidemic emergency throughout the national territory*), informally known as the "#iorestoacasa Decree" extending containment measures and provisions of the previous decree to the entire nation—the Bergamo area recorded 1472 Covid-19 infections, becoming the province with the greatest count of infected persons in Italy. For more details on the trend of infections in Italy see: http://opendatadpc.maps.arcgis.com/apps/opsdashboard/index.html#/b0c68bce2cce478eaac82fe38d4138b1 (accessed on 24 February 2021).

[j]Map processing was carried out on data provided by the Italian Higher Health Institute (*Istituto Superiore di Sanità*) concerning the number of infected individuals by municipality, updated to April 14, 2020 (https://www.epicentro.iss.it/en/coronavirus/sars-cov-2-integrated-surveillance-data). After that date, no further data was officially released on the number of patients on a municipal scale. Areas color-coded in white refer to a number of infected persons equal to zero; the same color-coding also includes municipalities for which the number of infected persons remains unspecified.

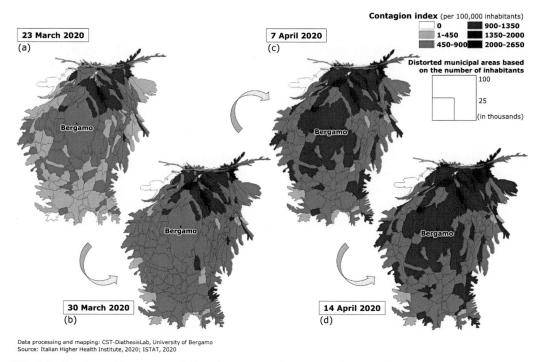

FIG. 1.9   Province of Bergamo: evolution of contagion relative to resident population.

covered by mountain territories—whose municipalities lie to the north along the Bergamo valleys—are visually shrunk to reflect their low population number. The rate of people who contracted the Covid-19 virus, on the other hand, was charted along a *range* and shown via color-coding. The different shades of red used in the map visualize contagion probability for a given person within a specific area while also designating such area as a "danger zone."

The Fig. 1.9 sequence outlines a progressive increase in the number of infections over time: at first viral propagation (a) affects mainly the Seriana Valley, which together with some municipalities of the Brembana Valley and Imagna Valley, records the highest contagion rates in the province. The city of Bergamo, with several municipalities located further south and the rest of the valleys, has a uniform and a moderately low rate of contagion; while the lower Bergamo area has a relatively weaker infection spread. As we move to map (b), we note that the outbreak has extended to the entire plain area of the southern province, in which nearly all the municipalities record a uniform incidence of infection which exceeds 450 cases per 100,000 inhabitants. In addition, there is an infection increase in the northern municipalities, along the Bergamo valleys. In the third map (c), which shows data 1 week later, further viral intensification is recorded—in particular in the central area of Bergamo—which impacts Bergamo, its immediate surroundings, and the crown of municipalities that encircles it. In addition, infection escalates even in mountain municipalities, located north of Bergamo, some of which report a very high contagion index, reaching over 1350 infected persons per 100,000 inhabitants. Finally, the last map (d)—dated April 14, 2020—shows further epidemic aggravation throughout the province, particularly in the Seriana Valley,[k] which is besieged by contagion and records the absolute highest infection rates in the Bergamo area.

Cartographic representation forcefully indicates that the Seriana Valley (Bergamo's Covid-19 hotbed), and the municipalities of Nembro, Albino and Alzano Lombardo, located in the southern area of the valley, are among the first provincial areas to have recorded a high number of infected people, even though the epidemic spread slowed down in time. Conversely, in areas reached by the Covid-19 epidemic at a later stage, as in the most central and densely populated municipalities of the Bergamo province (including Bergamo, Seriate, Dalmine and Treviglio), the number of people affected by the virus progressively increased.[l] This corroborates the hypothesis whereby viral propagation occurred first via reticularity, due to dense contacts across periurban areas where Covid-19 hotspots flared up (Lévy, 2020); and subsequently via proximity contacts, as the virus swiftly spread throughout large housing clusters.

Epidemic spread in the province of Bergamo was analyzed in relation to the population distribution and to infection data. This has yielded further information on Covid-19 diffusion in the territories under examination.

Alternative cartographic techniques made it possible to collate different datasets about Covid-19 in order to assess the actual extent of the epidemic. Surface extrusion was used to distort the surfaces of municipalities, with a view to creating three-dimensional shapes that convey the absolute number of infected people for each area of the province of Bergamo at different times. Color-coding was used instead to visualize contagion rates, with tonal ranges reflecting information on viral incidence for each municipality. A visualization of this kind allows researchers to retrieve extremely valuable information which, unlike purely quantitative data, can convey the social significance of contagion.

The Fig. 1.10 map indicates that not all the municipalities most affected by the virus in absolute numbers also have a high contagion rate. These areas mainly coincide with the Seriana Valley territory and are close to Bergamo's infection hotbed, graphically rendered on the map with a deflagration symbol. The map shows that the greatest number of patients in absolute terms are to be ascribed to the city of Bergamo and the towns of Seriate, Dalmine, Treviglio, Albino and Nembro, where viral incidence is considerable and reaches elevated contagion levels.[m] Even though municipalities north of Bergamo, mainly located in the Seriana Valley, show mild *extrusion*—which indicates a rather moderate absolute number of infected people—they have a very high contagion index, which tells us that the risk of contracting the disease in those areas is equally elevated.

[k]The municipality of Songavazzo, located in the northern area of the Seriana Valley, deserves attention, because it shows a high incidence of contagion, exceeding 2000 infections per 100,000 inhabitants. The data, however, must be related to the resident municipal population, which counts only 722 inhabitants.

[l]This is confirmed by regional reports issued in weekly monitoring by the Italian Ministry of Health: http://www.salute.gov.it/portale/nuovocoronavirus/dettaglioNotizieNuovoCoronavirus.jsp?lingua=italiano&id=5025 (accessed 24 February 2021).

[m]As of April 14, 2020, date of the last survey on a municipal scale, the municipal distribution of absolute infections for the Bergamo province shows that the city of Bergamo is the most affected (with 1332 total infections), followed by Seriate (255), Dalmine (242), Treviglio (235), Albino (223) and Nembro (220) (data: Higher Health Institute, https://www.epicentro.iss.it/en/coronavirus/sars-cov-2-integrated-surveillance-data).

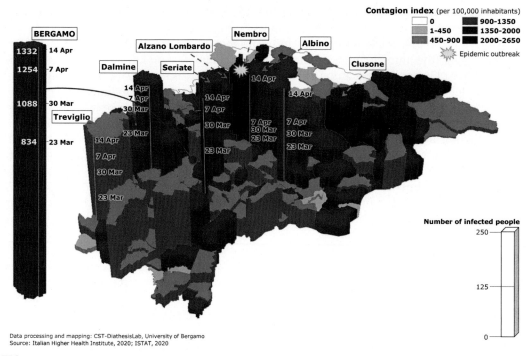

FIG. 1.10   Province of Bergamo: distribution of contagion from 24 February to 14 April 2020.

What cartographic representation leads us to conjecture is that in order to understand the Covid-19 epidemic in the province of Bergamo we need to retrieve data about its socio-territorial context. Once cross-referenced to absolute contagion data, such information projects an epidemic scenario quite unlike the one provided by mere geolocation of the patient counts for specific areas. Cartography makes it possible to transcend sheer geolocation data and gain a more articulate understanding of the actual dynamics of infection.

1.4.3 The Seriana Valley hotspot

The first case of Covid-19 infection in the Bergamo area was officially confirmed on 23 February 2020 at the Hospital of Alzano Lombardo, a municipality located in the Seriana Valley.[n] The valley, which is part of the Orobie Alps, is located north-east of Bergamo. The valley territory winds along the course of the Serio river, with population and settlement clusters that are quite diverse in their administrative and functional setups. The Lower Seriana Valley (or Lower Valley) includes 18 municipalities that comprise 71.7% of the entire valley's population: the towns of Albino (which has over 18,000 inhabitants), Alzano Lombardo (with about 14,000) and Nembro (about 12,000) are the largest clusters, located on the valley floor. They were severely affected by the epidemic. The Upper Valley (which comprises

---

[n] For a more detailed reconstruction of the events that led the area from the Seriana Valley to be among the first Covid-19 hotbeds in Italy, see Chapter 5 of this volume.

20 municipalities) hosts fewer residents (28.3% of the valley's population) but includes three larger centers: the town of Clusone and the towns of Castione della Presolana and Rovetta, where infections were still high, albeit with a lower incidence. Overall, the Seriana Valley therefore delineates a strongly urbanized and dense territory to the south, where a conurbation extends; and a series of smaller urban clusters along the slopes and in the lateral uplands, which however are highly developed in the textile and mechanical sectors.

To shed light on why Covid-19 swept so swiftly and lethally across this territory, we have used geographical tools for infection data modeling, and, at the same time, we took on a socio-territorial perspective. The settlement configuration of the Seriana Valley was represented cartographically (Fig. 1.11), conveying the morphological features of the region via tridimensionality, which traces the valley course.[o] Color-coding was used instead to mark the spread of infection recorded in the hotbed area in April 2020,[p] namely in Albino and Nembro (located in the Lower Seriana Valley). These two towns recorded the greatest number of Covid-19 patients, with the third town of Alzano Lombardo trailing right behind. With the exception of the town of Clusone, the Upper Valley seems instead to show a lower distribution of infection. This could be ascribed to the lower number of inhabitants[q] and to the fact that this specific area extends over an indented plateau, where a settlement configuration of small towns and many scattered houses could have favored distancing and a local decrease in relational networks. Local territory conditions did not favor the establishment of productive activities or services, which entailed a decrease in the resident population. What matters in this area is, rather, mobility and the commuting of people for study or work.[r]

Cartography about school mobility shows the fluxes of commuters who move daily towards the main school municipalities along the Seriana Valley (Fig. 1.12). Schools located in the towns inside the Lower Valley conurbation emerge as attractors not only for students who reside in the Seriana Valley, but also for those who commute to the Lower Valley for study reasons, even though they are domiciled in Bergamo, in the municipalities adjacent to the valley entrance or in the lower Bergamo plain. This mobility pattern corresponds to 92% of school flows and is in strong contrast with commute flow data for the Upper Valley, which records only 8% of movements for study reasons.[s] Data on work commuting is also

[o] The basemap makes it possible to represent the territory to be analyzed from a physical point of view, while simultaneously retrieving altimetric information. Landscape cartography seems particularly important in this regard, because perspectival figuration can successfully be made to convey the values and knowledge sets that a given society attributes to its territory, bringing out its complex features (Casti, 2015, pp. 86–90).

[p] In particular, this representational method enables cartographers to limit figuration of the actual infection spread for inhabited centers, instead of uniformly scattering data throughout the whole municipal surface. Data provided by the Higher Health Institute pertain infected by municipalities up to the date of 14 April 2020 (https://www.epicentro.iss.it/en/coronavirus/sars-cov-2-integrated-surveillance-data).

[q] It should be remembered that the Lower Valley conurbation comprises 100,000 inhabitants; the upper valley instead is home to nearly 40,000 residents (http://demo.istat.it/index_e.php).

[r] The vibrant fabric of the Lower Valley derives from the prosperity of the local industry sector, which however facilitated the propagation of Covid-19 during the epidemic. For more details on the role of mobility in the spread of the Covid-19 virus, see Chapter 3 of this volume.

[s] Schools in the northern valley area attract students who mainly reside in the Upper Seriana Valley. To a lesser extent, they attract commuters for study from the neighboring valleys (for example, the Val di Scalve) and from some neighboring municipalities further south.

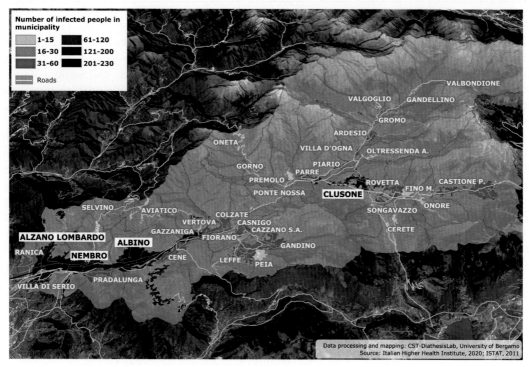

FIG. 1.11   Seriana Valley: distribution of the Covid-19 infection (absolute data) from 24 February to 14 April 2020.

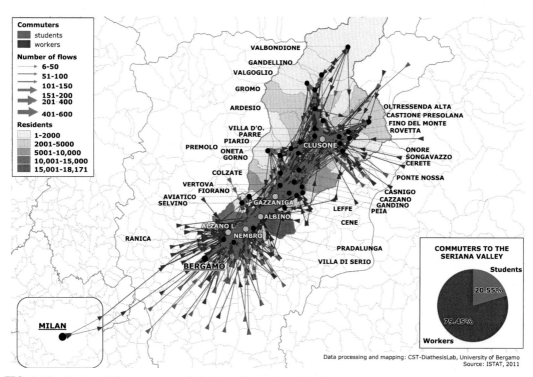

FIG. 1.12   Seriana Valley commutes: daily flows for study and work.

substantial, especially in the Lower Valley.[t] High mobility flows are determined by the highest number of local companies and industries, which employ a large workforce.[u] Cartography clearly visualizes marked commute flows along the Seriana Valley (almost 80% workers and 20% students). It also clearly records that workers[v] residing in the most isolated areas along the valley slopes or in areas adjacent to the valley usually commute to the valley floor, towards the most important urban clusters which host company headquarters. Furthermore, cartography highlights incoming commute flows and, more precisely, commutes from the municipalities of the plain areas beyond the valley and from the city of Bergamo.[w]

The linchpin of this dynamic network rests with the very municipalities that may be identified as the hotspots of the Bergamo outbreak, since these are obligatory points along the passage to the northern provincial area, given that the Valley's morphology allows for the presence of only one main central communication corridor. It is therefore reasonable to suppose that daily and constant flows of inhabitants, in other words intense commuting understood as an indicator of individual movement for study or work reasons between different territories, may have favored overcrowding thereby leading to intensification of Covid-19 contagion, especially at peak times in daily public transport commutes.[x] The intricate pattern of territorial interconnections is likely to have aggravated contagion risks. This, combined with other competing epidemic determinants such as a centralized healthcare model instead of a widespread and targeted local network of care, undoubtedly exacerbated the epidemic development across the Seriana Valley.[y]

[t]Lower Valley commutes cover a daily flow of about 26,000 workers, accounting for 81% of work commutes. Work commutes towards the Upper Valley correspond only to 19% of the flows.

[u]Suffice it to say that Nembro and Alzano, two towns located at the mouth of the valley, alone host 376 companies with nearly 4000 employees. The Seriana Valley thus features a solid manufacturing and production sector, closely rooted in the local territory and tied to other areas both on a national and an international scale. Please see: https://www.bg.camcom.it/lang/en (accessed on 24 February 2021).

[v]With reference to the active population residing in the municipalities of the Seriana Valley, we need to take into account the dichotomy between the populous Lower Valley, which has a higher workforce numbers, and the Upper Valley, where a smaller number of people of working age resides. Active population is classified by the Italian Office for National Statistics (ISTAT) according to Eurostat definitions and is divided into employed and unemployed (or job seekers), aged over 15 years. See https://www.istat.it/en/methods-and-tools/glossary (accessed on 24 February 2021).

[w]We should also take into account the service of the so-called "Tram delle Valli" (or T1), a railway train that runs between the urban center of Bergamo and the municipality of Albino. It is an interurban light rail line serving the city of Bergamo and the Lower Seriana Valley.

[x]As regards the relation between infection spread and mobility, see the paper titled *Annual report 2020. The Italian scenario* envisaged by the Office for National Statistics specifies that "one of the determinants of Covid-19 localization and of variable epidemic spread throughout the territory lies with work mobility. Mortality analysis with reference to the geography of local labor systems (SLL), allows for a description of Covid-19 epidemic impact based on a characterization of territory that takes note of daily urban systems, i.e., those places where people work and have most of their social and economic relations" (ISTAT, 2020, p. 86).

[y]It is widely agreed that shortcomings in the local healthcare system (in particular, the Alzano Lombardo hospital) and the failure to quickly establish a so-called "red zone" of quarantine in the Municipalities of Nembro and Alzano Lombardo played a major role in the spread of Covid-19 across the Seriana Valley and, in general, throughout the provincial territory. The Seriana Valley case has been legally defined as a criminal law issue and is currently being investigated by the Italian Judiciary (Magistratura).

1.4.4 Conclusions

The Seriana Valley presents a range of territorial variables that are likely to have facilitated the spread of Covid-19, which reached Italy and specifically Lombardy after making its appearance in China. Covid-19 found the territory ill-prepared to cope with an unexpectedly swift diffusion, and researchers were from the start left wondering why hotbeds had suddenly flared up across Lombardy and, most vehemently, in the province of Bergamo. Analysis of the morphological and social features that characterize the Bergamo area, and especially the Seriana Valley, has pinpointed a set of factors that are most likely to have facilitated the rapid epidemic spread.

First is the fact that the Bergamo area partakes of the polycentric conurbation typical of Lombardy, a region which may be said to have a *unique* settlement configuration. This urbanized territorial setup has crucial implications in times of pandemic, because it relies on a network of movements or commutes, which facilitate viral propagation by reticularity, and is characterized by high population density, which consequently leads to a potential increase in the possibility of spreading the disease by proximity. With particular regard to the Seriana Valley, it should be noted that this territory is marked by the presence of highly developed and flourishing industries, which involve a major commute and movement flows along one single infrastructural route seamlessly extended to the Lower Valley and responsible for high levels of pollution.[z] The same territory has a high population density, which enabled Covid-19 to sweep through the valley by reticularity and by proximity, making the valley a major hotbed of viral propagation, which then reached and impacted the entire province of Bergamo.

The guiding premise of our research, namely that contemporary mobile and urbanized living may facilitate the speed and severity of Covid-19 contagion, finds corroboration in the analysis of the specific case of the Seriana Valley. This case thus confirms the hypothesis whereby the socio-territorial features of areas most forcefully affected by viral outbreaks played a crucial role in the propagation of the pandemic.

[z] Among the main air pollutants is nitrogen dioxide (NO_2), which is considered a secondary pollutant since it is not the direct result of exhausts or industrial fumes, but generally derives from conversion of nitrogen monoxide (NO) via oxidation. Nitrogen dioxide gas is harmful to human health: it can cause acute respiratory dysfunction and bronchial reactivity (irritation of the mucous membranes) and increase the risk of contracting tumors. Similarly to the Bergamo area, the Seriana Valley is strongly affected by this form of air pollution; however, in 2019, average annual levels allowed by law were exceeded only in one municipality (Ponte Nossa). For more information, see the ARPA Lombardia website (https://www.arpalombardia.it/) and Chapter 4.1 of this volume.

References

Casti, E., 2000. Reality as Representation. The Semiotics of Cartography and the Generation of Meaning. Bergamo University Press, Sestante, Bergamo.

Casti, E., 2015. Reflexive Cartography. A New Perspective on Mapping. Elsevier, Amsterdam.

Casti, E., 2020. Geografia a 'vele spiegate': analisi territoriale e mapping riflessivo sul Covid-19 in Italia. Documenti Geografici 1, 61–83.

De Angelis, I., Trotta, E., Trecci, F., Romagnoli, T., 2019. Bergamo e i suoi territori. In: Motore di investimenti in Real Estate e Infrastrutture per lo sviluppo sociale e la crescita economica lombarda. Associazione Nazionale Costruttori Edili.

Ferlinghetti, R., 2015. S-low Bergamo: green agricultural spaces in urban evolution. In: Casti, E., Burini, F. (Eds.), Centrality of Territories. Verso la rigenerazione di Bergamo in un network europeo. Sestante Edizioni, Bergamo, pp. 111–140.

ISTAT, 2020. Rapporto annuale 2020. La situazione del Paese. accessed October 2020 from https://www.istat.it/it/archivio/244848.

Further reading

Fraser, T.D.R., 2019. Cybercartography revisited. In: Fraser, T.D.R., Anonby, E., Murasugi, K. (Eds.), Further Developments in the Theory and Practice of Cybercartography. Elsevier, Amsterdam, pp. 3–23.

ISTAT, Istituto Superiore di Sanità (Ed.), 2020. Impatto dell'epidemia Covid-19 sulla mortalità totale della popolazione residente. Primo trimestre 2020. (Ed.), accessed October 2020 from https://www.epicentro.iss.it/coronavirus/pdf/Rapporto_Istat_ISS.pdf.

Lévy, J., 2020. L'humanité habite le Covid-19. AOC. Analyse, Opinion, Critique. accessed October 2020 from https://aoc.media/analyse/2020/03/25/lhumanite-habite-le-covid-19/.

Lévy, J. (Ed.), 2008. L'invention du monde. Une géographie de la mondialisation. Presses de Sciences Po, Paris.

Lussault, M., 2020a. Le Monde du virus—retourn sur l'eprouve de confinement. AOC. Analyse, Opinion, Critique. accessed October 2020 from https://aoc.media/analyse/2020/05/10/le-monde-du-virus-retour-sur-lepreuve-du-confinement/.

Lussault, M., 2020b. Chroniques de géo' virale. Ecole urbaine de Lyon/Editions deux-cent-cinq, Lione.

Mavagrani, A., 2020. Tracking Covid-19 in Europe: infodemiology approach. JMIR Public Health Surveill. 6 (2). accessed October 2020 from https://publichealth.jmir.org/2020/2/e18941/.

Chapter 2. Mortality and severity of contagion

Estimation of mortality and severity of the Covid-19 epidemic in Italy

Ilia Negri and Marcella Mazzoleni

2.1.1 Introduction

Starting from the first case recorded, information on the spread of Covid-19 in Italy is communicated daily by the institutions. The number of deaths related to Covid-19 is provided by the ISS (Istituto Superiore di Sanità—Italian Higher Health Institute) but only at the regional level and by age at the national level. At present, information on the age of deceased persons at regional level has not yet been given. Also, during the most critical phase of contagion, several people died without a swab test that would have established whether they had contracted the virus. For these reasons, we decided to analyze death rates using ISTAT (Italian Office of Statistics) mortality tables.

In October 2020, ISTAT published a note (ISTAT, 2020) which reported up-to-date data on deaths from all causes in Italy by municipality and by age in the first 8 months of the year from 2015 to 2020. While in July 2020 ISTAT (ISTAT, Istituto Superiore di Sanità, 2020c) table which addressed almost all of the Italian territory, 87% of the 7904 Italian municipalities and 86.4% of the inhabitants, the updated report they covered the entire territory and all the Italian population. ISTAT issued a first report on mortality in May 2020 (ISTAT, Istituto Superiore di Sanità, 2020a), followed by a second report (ISTAT, Istituto Superiore di Sanità, 2020b) published in June 2020.

In the last published report, ISTAT (ISTAT, 2020) provided data relating to all the subjects who died daily in Italy by municipality and by age in the first 8 months of the year from 2015 to 2020: these are the data we consider in this study, along with data provided by the ISS on Covid-19-related deaths established via swab tests.

We propose to estimate mortality by region and by age group due to both Covid-19 and causes attributable to Covid-19. This estimate is compared to deaths from other causes, or mortality that we would have expected to observe without the impact of the disease.

The aim is to understand the impact that the epidemic had in the different regions and in the different age groups. In fact, by comparing mortality rates in 2020 and in previous years (from 2015 to 2019) with the official number of Covid-19-related deaths, it is possible to observe an excess of deaths in some geographical areas and for some age groups.

ISTAT (ISTAT, Istituto Superiore di Sanità, 2020a) states that this excess may be attributed to three causes: additional mortality due to Covid-19 (deaths in which a swab test was not performed); indirect mortality associated with Covid-19 (deaths from organ dysfunction for the heart or kidneys caused by disease triggered by the virus in untested people) and, finally, an indirect mortality rate unrelated to the virus but caused by the meltdown of the public health system and the reluctance of patients to undergo hospitalization in the most affected areas.

ISTAT mortality tables, and other official mortality data disclosed by the ISS, have been widely addressed in the literature and the severity of the Italian situation emerges clearly in all the articles.

Along these lines, Modi et al. (2020) performed counterfactual analysis of the historical series on the 2020 mortality data using control data provided by ISTAT on the previous 5 years. They found that excess mortality significantly exceeds official Covid-19 death rates. They suggest that there is a large amount of people, mostly elderly, missing from official mortality statistics. They estimate the case mortality rate (CFR), the death rate from infection (IFR), the population death rate (PFR), and the infection rate (IR). Finally, they note that the estimation of IFR and IR is very difficult due to uncertainties in the official Covid-19 death rate.

Buonanno et al. (2020) provide new results on the misreported mortality level for Lombardy and the Bergamo province using both official and original data sources. By combining official statistics, retrospective data, and original data (i.e., obituaries and death notices) they compare the official number of deaths reported in March 2020 with those reported in the same month in the previous year in a sample of Lombard municipalities.

Marino and Musolino (2020) focus on ISTAT mortality tables to compare official and "hidden" Covid-19 mortality. They indicate that several people with the disease died at home without being tested. They observe a significant difference between official and "hidden" mortality in Lombardy, especially in some of its provinces such as Bergamo and Brescia, while this difference is smaller for the Center and even lower for Southern Italy.

Bucci et al. (2020) underline the fact that the deaths may be divided into direct and indirect related to Covid-19, where indirect deaths are due to an overload of the health system. They show that the official figures underestimate the number of deaths, especially in the most affected areas. In their study, they focus on Lombardy, Liguria and Emilia-Romagna using an older dataset and with fewer data than the present model. They use gender imbalance to estimate disease-related death. They separate expected deaths in a normal situation, with deaths related either directly or indirectly to Covid-19.

Murgante et al. (2020) estimate the number of deaths expected in the Italian provinces from the sum for each age of the population of a specific age in each province multiplied by the national mortality rate for that specific age. The standardized death rate is then estimated by comparing the number of events recorded in each province with the respective number of expected events.

Along these lines, our goal is to assess mortality attributable to Covid-19 and to other causes by charting age-specific estimates for each region. We therefore propose to estimate mortality rate using both official data sources: Covid-19-related mortality in all regions provided by the ISS and overall mortality (therefore both from Covid-19 and from other causes) provided by ISTAT. For the purposes of our research, we will consider the population death rate provided by ISTAT as of 1 January 2019.

Analyzing and comparing the situation in Italy and in other countries, several authors have laid out a range of hypotheses to account for different mortality levels.

Yuan et al. (2020) focus on transmissibility and mortality of Covid-19 in Europe and report real-time, actual reproduction numbers and case fatality rates in Europe by using data from the World Health Organization website.

Fanelli and Piazza (2020) analyze the temporal dynamics of the Covid-19 outbreak in China, Italy, and France, within the time frame 22/01–15/03/2020, the initial phase in Italy and France. Analysis of the same data within a simple susceptible infected-recovered-deaths model indicates that the recovery rate does not seem to depend on the country, while the infection and death rate show a more marked variability.

Onder et al. (2020) indicate that the fatality rate in the Italian population, based on data up to March 17, 2020, was 7.2%. This rate is higher than the one recorded in other countries and may be related to three factors. The demographics of the Italian population differ from other countries, so the older age distribution in Italy may partly explain Italy's higher case-fatality rate compared to other countries. A second possible explanation may have to do with the way in which Covid-19-related deaths are identified in Italy. Covid-19-related deaths as those occurring in patients who test positive for Covid-19, regardless of pre-existing diseases that may have caused death. A third possible explanation are the differing strategies used for Covid-19 testing. After an initial, extensive testing strategy of both symptomatic and asymptomatic contacts of infected patients in a very early phase of the epidemic, on February 25, 2020, the Italian Ministry of Health issued more stringent testing policies, mandating testing only for patients with more severe clinical symptoms who were suspected of having Covid-19 and required hospitalization.[a]

Medford and Trias-Llimós (2020) explore differences in age distribution of deaths from Covid-19 among European countries by a cross-country comparison of age distribution and put forward some reasons for potential differences. They analyze the situation in France, Italy, the Netherlands, and Spain with a cross-country comparison of age-patterns in observed Covid-19 death counts and their counterfactual distribution adjusted by the age-structure of Italy. Italy has the oldest population in Europe, but proportionately fewer deaths from Covid-19 at older ages are found in Italy than in either Spain, France, or the Netherlands.

This chapter is laid out as follows. The first section features an analysis of mortality in March 2020 in Italy; the second provides estimates for deaths directly or indirectly attributable to Covid-19; the third reports previous estimates divided by age; and the last section is devoted to conclusions.

---

[a]For more details, see Chapter 5 of this volume.

2.1.2 Mortality in Italy in March 2020

According to ISTAT mortality tables (ISTAT, 2020), in March 2020 85,786 people died in Italy against an average of 58,265 over the previous 5 years. As shown in Table 2.1, the region where the death toll was higher than the average number of deaths recorded in the previous 5 years for the same period is Lombardy. Virtually one third of the people who died in Italy in March 2020 were in Lombardy. While some regions (such as Aosta Valley, Molise, or Basilicata) show very low absolute numbers, it would be erroneous to conclude that Covid-19 had a minor impact in them.

If we analyze the percentage change in deaths in March 2020 from Table 2.1, and compare it to the average death toll in March over the previous 5 years, we notice that Lombardy presents the largest variation with an increase of almost 191 deaths. It may be concluded that the

TABLE 2.1   Total deaths in March 2020; average death rate over the previous 5 years in the same period; and percentage change (calculated from ISTAT mortality data).

Total deaths	2020	2015–2019	% variation
Piedmont	7240	4740	52.74%
Aosta Valley	205	134	52.99%
Liguria	3024	1970	53.50%
Lombardy	25,560	8778	191.18%
Trentino-South Tyrol	1415	872	62.27%
Veneto	5413	4457	21.45%
Friuli-Venezia Giulia	1519	1353	12.27%
Emilia-Romagna	7748	4582	69.10%
Marche	2328	1620	43.70%
Tuscany	4511	3998	12.83%
Umbria	1043	963	8.31%
Lazio	5384	5245	2.65%
Campania	5083	5026	1.13%
Abruzzo	1557	1372	13.48%
Molise	371	361	2.77%
Apulia	4011	3599	11.45%
Basilicata	571	598	−4.52%
Calabria	1972	1907	3.41%
Sicily	5117	5149	−0.62%
Sardinia	1714	1541	11.23%
ITALY	85,786	58,265	47.23%

deaths in Lombardy almost tripled: for every 100 deaths in March last year, 291 were observed this year (2020). That is followed by Emilia-Romagna, Trentino-South Tyrol, Liguria, Aosta Valley, and Piedmont, where the percentage change is greater than the percentage change recorded in Italy, which is 47%. The regions where the change was lower, if not negative, are Basilicata, Sicily, Campania, Molise, and Calabria.

We may conclude that in Aosta Valley, although the absolute number of deaths is low, the impact of Covid-19 was greater than in other regions which show higher absolute numbers.

This fact is also confirmed by a mortality rate calculation. The death rate for each region in March 2020 is calculated as the ratio between the deaths in March 2020 and the population residing in that region on January 1, 2019. The number obtained is multiplied by 100,000 to obtain the number of deaths for each 100,000 inhabitants. Table 2.2 shows the mortality rates in March for each region calculated in 2020, compared to the same rate calculated on

TABLE 2.2   Mortality rate for each 100,000 inhabitants in March 2020 and average deaths over the previous 5 years in the same period (calculated from ISTAT mortality data).

Mortality rate each 100,000 inhabitants	2020	2015–2019
Piedmont	166	109
Aosta Valley	82	53
Liguria	195	127
Lombardy	254	87
Trentino-South Tyrol	132	81
Veneto	110	91
Friuli-Venezia Giulia	125	111
Emilia-Romagna	174	103
Marche	153	106
Tuscany	121	107
Umbria	118	109
Lazio	92	89
Campania	88	87
Abruzzo	119	105
Molise	121	118
Apulia	100	89
Basilicata	101	106
Calabria	101	98
Sicily	102	103
Sardinia	105	94
ITALY	*142*	*96*

the average number of deaths in March over the 5 years preceding 2020, again divided by the population of 1st January 2019.

In 2020 the death rate in Italy rose noticeably to 142 per 100,000 inhabitants from 96 in previous years. Also, the increase that occurred in all regions has a very different dimension from region to region, placing Lombardy as the region with the highest mortality rate, equal to 254 deaths per 100,000 inhabitants.[b]

As regards the mortality rate, within the Lombardy region the situation is very different by province and this will be explored in Chapter 2.2.

2.1.3 Mortality estimation for Covid-19

Starting from these March mortality data and the data reported by the ISS on deaths from Covid-19 in the regions, we decided to estimate the impact that Covid-19 had on mortality in the various regions of Italy. Our aim is to estimate the number of deaths from all causes attributable to Covid-19. We call it "number of deaths due to Covid+", where the + stands not only for deaths due to Covid-19, but also for the other three causes related to the presence of the virus in the Italian territory. We denote this number with C_{re}^+, where r stands for the geographic area and e denotes the age group. The estimate is done before marginally in each region of Italy. In this phase the estimate of both $C_{r.}^+$ and $C_{.e}^+$ is calculated, respectively the total number of deaths in the area r and the total number of deaths in the age group e. Then, the value of C_{re}^+ is estimated for any area r and for any age group e. Let D_{re} denote the number of deaths in area r and in the age group e and G_{re} the number of deaths due to other causes not attributable to Covid-19 in area r and in the age group e. We have the following equality: $D_{re} = G_{re} + C_{re}^+$. Let C_{re}^{ISS} be the number of official deaths from Covid-19 in area r and age e. This last number is not given by the ISS. Indeed, $C_{r.}^{ISS}$ and $C_{.e}^{ISS}$, respectively, the total deaths for Covid-19 in the area r and in the age class e, are known and provided by ISS but, unfortunately, they underestimate the deaths for Covid-19 in some areas and for some age's classes.

Starting from the marginal, the deaths due to Covid+ for each age e is the difference between the total deaths for that age $D_{.e}$ and $G_{.e}$. This last number is estimated by the mean of deaths in the previous years for the age class e. Accordingly, for each age e:

$$C_{.e}^+ = \begin{cases} D_{.e} - G_{.e}, \text{ if } D_{.e} - G_{.e} > C_{.e}^{ISS} \\ C_{.e}^{ISS}, \text{ otherwise} \end{cases}$$

where $C_{.e}^+$ are the deaths for Covid+ for the age e.

If we move on to an analysis of each area r, we find that the total deaths for Covid+ for each area r is the difference between the deaths in that area $D_{r.}$ and $G_{r.}$. This last number is estimated by the mean of deaths in the previous years in the area r. Accordingly, for each area r we define:

---

[b] These results are coherent with the analyses on the spread of the Covid-19 infection developed in Chapter 1 of this volume.

$$C_{r.}^+ = \begin{cases} D_{r.} - G_{r.}, \ if \ D_{r.} - G_{r.} > C_{r.}^{ISS} \\ \qquad C_{r.}^{ISS}, \ otherwise \end{cases}$$

where $C_{r.}^+$ are the deaths for Covid$+$ in area r.

Once the marginals have been estimated it is possible, at a first stage, to estimate the C_{re}^+, which indicates the deaths for Covid$+$ in the area r and for age e, such as:

$$C_{re}^+ = \begin{cases} D_{re} - G_{re}, \ if \ D_{re} - G_{re} > 0 \\ \qquad 0, \ otherwise \end{cases}$$

Accordingly, C_{re}^+ for each age e and for each area r is the difference between deaths D_{re} and G_{re}. This last number is estimated with the mean of deaths in the previous years in the area r and for the age class e.

C_{re}^+ must be adjusted according to the estimated marginal. For an age class $e \ \sum_{r=1}^R C_{re}^+ < C_{.e}^+$ or for a geographical area $\sum_{e=1}^E C_{re}^+ < C_{r.}^+$, we have an underestimation of deaths for Covid-19. Therefore, if the sum for each area or for each age is not greater or equal to the total deaths for Covid$+$ for that area or for that age class just estimated (our constraints), the estimate for that age class and for that area must be updated.

Starting from the total deaths for Covid$+$ for each age e, if $\sum_{r=1}^R C_{re}^+ < C_{.e}^+$, the new estimate indicated with $C_{re}^{+\prime}$ becomes for any r:

$$C_{re}^{+\prime} = C_{re}^+ + \left(C_{.e}^+ - \sum_{r=1}^R C_{er}^+ \right) \frac{C_{r.}^+}{C_{..}^+}$$

where $C_{..}^+$ indicated the sum of deaths for Covid$+$ in Italy. In accordance with the formulation the excess deaths for these ages are divided by the proportion of deaths for Covid$+$ in that area over the total deaths for Covid$+$ in Italy. This formula is used for each age except for the last one E, the oldest group, where in each area, it is possible to observe an excess of deaths. Then the value for the last age is obtained by the difference: $C_{rE}^{+\prime} = C_{r.}^+ - \sum_{e=1}^{E-1} C_{er}^{+\prime}$, so the total deaths by area does not change.

Concerning the total deaths for Covid$+$ in each area r, if $\sum_{e=1}^E C_{re}^+ < C_{r.}^+$ we estimate the deaths for Covid$+$ for each age class e starting from the oldest ages using this formulation, for any e:

$$C_{re}^{+\prime\prime} = C_{re}^{+\prime} + \left(C_{r.}^+ - \sum_{e=1}^E C_{re}^{+\prime} \right) * \left(\frac{C_{.e}^{ISS}}{P_{.e}} \right) * P_{re}$$

where P_{re} indicates the inhabitants in the area r for the age e and $P_{.e} = \sum_{r=1}^R P_{re}$ indicates the Italian inhabitants for the age e and we stop to update this value when $\sum_{e=1}^E C_{re}^+ = C_{r.}^+$. According to this formulation, we divide the excess of deaths between the age's classes starting from the oldest classes (the most affected by Covid-19) using the Italian mortality rate multiplied by the inhabitants in that area for that age class e.

Once the deaths for Covid$+$ are estimated for each area r and for each age e, the sum of deaths for each age $C_{.e}^{+\prime\prime}$ is updated. This value will be greater than $C_{.e}^+$ as we add new deaths in the ages starting from the oldest subjects, the most affected.

Lastly, the values G_{re} are updated, as D_{re} is fixed, we have just estimated C_{re}^+ and, accordingly, we obtain by difference the new values of $G_{re} = D_{re} - C_{re}^+{}''$.

In this section we consider the mortality data in each region of Italy in March 2020. The death rate for Covid+ is then calculated by dividing the estimated deaths, obtained using the above method, by the population residing in that region on 1st January 2019.

In Table 2.3 and Fig. 2.1, the estimated mortality rate from direct and indirect causes of Covid-19 (Covid+) is compared with the death rate due only to Covid-19, according to

TABLE 2.3   The mortality rate for each 100,000 inhabitants estimated for Covid+ and mortality rate due only to Covid-19 and total deaths for Covid+ and Covid-19 (calculated from ISTAT mortality data).

	Mortality rate Covid+	Mortality rate Covid-19	Deaths for Covid+	Deaths for Covid-19	Covid+/ Covid-19
Piedmont	57	23	2501	1018	2.46
Aosta Valley	56	56	71	70	1.01
Liguria	68	24	1055	368	2.87
Lombardy	167	83	16,782	8339	2.01
Trentino-South Tyrol	51	26	543	281	1.93
Veneto	20	10	957	509	1.88
Friuli-Venezia Giulia	14	5	166	57	2.91
Emilia-Romagna	71	42	3166	1886	1.68
Marche	46	22	709	328	2.16
Tuscany	14	6	513	226	2.27
Umbria	9	4	80	37	2.16
Lazio	3	3	158	158	1.00
Campania	1	1	79	79	1.00
Abruzzo	14	5	185	64	2.89
Molise	3	1	10	4	2.50
Apulia	10	3	412	118	3.49
Basilicata	1	1	5	5	1.00
Calabria	3	1	65	18	3.61
Sicily	2	2	77	77	1.00
Sardinia	11	2	173	39	4.44
ITALY	*46*	*23*	*27,707*	*13,681*	*2.03*

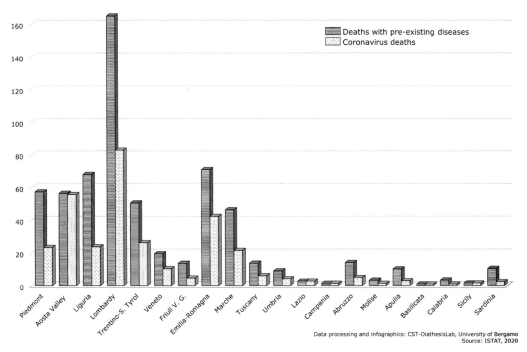

FIG. 2.1 Mortality rates in March 2020 for Covid-19 and Covid+ by region (estimates from ISTAT mortality data).

data released by the ISS. We can see that this value in Lombardy is almost double, which indicates that for every Covid-19 death certified by a swab test there is another death attributable to Covid-19 for direct or indirect reasons. Analyzing the relationship between Covid+ and Covid-19 it is possible to observe even higher values in other regions, such as in Sardinia, where this ratio is higher than 4 but linked to very low mortality values.

Fig. 2.2 shows an anamorphic map in which the surface is distorted (dilated or contracted) based on the number of inhabitants for each municipality on the 1st January 2019, which makes it possible to highlight the most populous areas. The number of deaths from Covid+ in March 2020 is color-coded, while circles are used to indicate the death rate for Covid+ per 100,000 inhabitants in each region. This reflexive cartography model—which presents localized and cross-referenced data—allows us to interpret mortality in its impact on population and on its regional distribution in a detailed and, at the same time, comparable view. This recovers the social impact of epidemic mortality and promotes a search for the possible causes of these differences in other socio-territorial data such as type of settlement, mobility, pollution or other.

An analysis of the regional level modified by the number of inhabitants shows that Lombardy, Lazio, and Campania are dilated in relation to the high population density of the metropolitan cities of Milan, Rome, and Naples that they enclose. The other regions are distorted with respect to their topography, because within them the municipalities are skewed according to the number of residents.

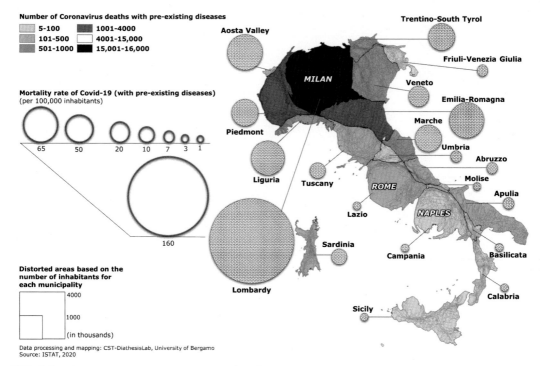

Data processing and mapping: CST-DiathesisLab, University of Bergamo
Source: ISTAT, 2020

FIG. 2.2   The number of deaths and mortality rate for Covid+ in March 2020.

Once again, the map indicates that Lombardy has the highest death rate from Covid+ (circle with the largest diameter) and the highest number of deaths (given by the gray color in the darkest shade). In Aosta Valley there is a high mortality rate value (marked by a larger circle) but a low number of deaths (color-coded in light gray). It means that the impact of the disease in this region with few inhabitants was high, for even though the absolute number of deaths is low, the death rate is high.

In other regions, such as Emilia-Romagna and Piedmont, we have a high number of deaths (color-coded in dark gray) but a lower mortality rate (circle with a smaller diameter than that of Lombardy). In Liguria, Veneto, and Marche there are several deaths (−coded in gray) but with different mortality rates, lower in Veneto, with a smaller circle, compared to the other two regions. Tuscany, Lazio, Abruzzo, Apulia, and Sardinia show fewer deaths (−coded in light gray) linked also to a low mortality rate (smaller circle).

2.1.4 Analysis of mortality by age

In this early stage of the epidemic, the spread of Covid-19 had a different impact on different age groups. To analyze this disparity, we estimated the mortality rate for each age group. Covid-19 impact on people under the age of 60 is virtually negligible compared to impact in older age groups. Consequently, Fig. 2.3 shows a map of Italy where basemap colors

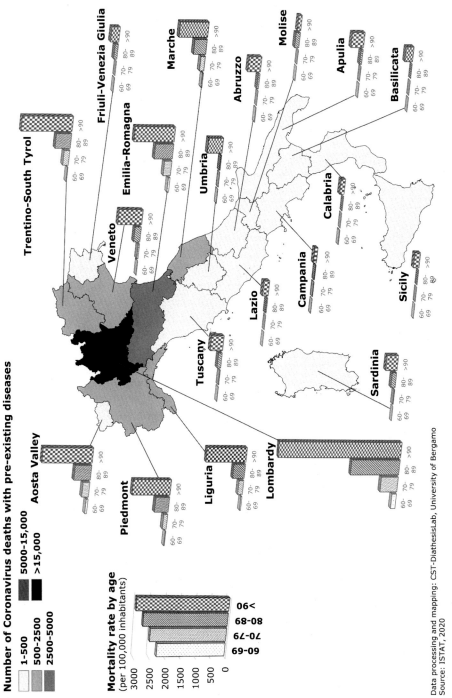

Data processing and mapping: CST-DiathesisLab, University of Bergamo
Source: ISTAT, 2020

FIG. 2.3   Mortality rate for Covid+ by region and by age in March 2020.

TABLE 2.4 Number of total deaths in March 2020, and deaths for Covid+ and Covid-19 by age (ISS and estimated from ISTAT mortality data).

Number of deaths	0–9	10–19	20–29	30–39	40–49	50–59	60–69	70–79	80–89	≥90	Total
Total	127	62	123	282	1023	3081	7141	18,165	34,353	21,429	85,786
Covid+	0	0	2	25	129	621	1916	6546	11,606	6862	27,707
Covid-19	0	0	2	25	92	409	1271	3724	4366	1051	10,940

(color-coded in gray) provide the total number of estimated deaths due to Covid+ in March 2020, while the death rate is reported in the bar plot for people aged 60 or older.

It may be observed that the older class (aged 90 or over) is the most affected. In fact, each region for this class presents the highest graph bar.

An analysis of the impact by age in each region indicates once again that Lombardy is the region most severely affected by the disease. Lombardy shows, in fact, the highest number of deaths (color-coded in black) and the highest mortality rate graph bar in each age group.

As already pointed out, Aosta Valley shows severe Covid-19 impact even with a low number of deaths: its area is color-coded as white to mark a low number of deaths, but graph bars are higher for each age group. Piedmont, Trentino-South Tyrol, Emilia-Romagna and Liguria show fairly high graph bars but a medium number of deaths (color-coded in light gray) with the exception of Emilia-Romagna (color-coded in dark gray), the second region for a number of deaths. For the Veneto and Marche regions, we observe a low number of deaths (light gray) and lower graph bars. In the other regions, there is a lower number of deaths (color-coded in white) and lower graph bars, except for the last bar that refer to the oldest age group. This last bar is higher in Sardinia, Tuscany, Umbria, Abruzzo, and Apulia than in other regions.

It should be emphasized that in the case of people over 70 years of age, Covid-19 is responsible for approximately one in three deaths, with 73,947 total deaths in March 2020 and 25,014 deaths from Covid+ (as shown in Table 2.4).

It should also be noted that the total number of deaths provided by ISS in Table 2.4 differs from the number given in Table 2.3. This last number is 2741 units greater than the first. Unfortunately, since the October 2020 ISTAT report (ISTAT, 2020) failed to include data on deaths for Covid-19 for age, we resorted to using the latest data provided by the ISS for 31/03/2020, available online.[c]

2.1.5 Conclusions

In this section we have considered data provided by ISTAT on mortality in Italy in the first quarter of 2020, focusing only on the month of March 2020, i.e., in the first and most severe phase in which Italy was faced with the emergence of the new SARS CoV-2 virus responsible for the Covid-19 disease.

[c]https://www.epicentro.iss.it/coronavirus/bollettino/Infografica_31marzo%20ITA.pdf.

Since there is a lack of information on age-related mortality by region, especially in regions, such as Lombardy, where the impact of Covid-19 was significant, we have estimated the mortality and mortality rate by combining data given by the ISS with those given by ISTAT for March 2020 and for the same month from 2015 to 2019, a year range used for comparison or control.

The conclusions are that the impact of the Covid-19 disease in Italy was severe, but Lombardy was the region that, in addition to the highest absolute numbers of deaths, revealed the highest rates of variation and the greatest mortality.

Similar conclusions may be drawn from official data, which confirm that Lombardy had the highest death rates. However, official data fail to quantify the impact this disease had on mortality. In fact, the mortality statistics compiled with our method indicate that the death toll was nearly double the one officially reported by the ISS, even though that was already high compared to mortality recorded for the same March period over the previous 5 years. This difference may be related to the fact that a very large number of deaths were not certified as Covid-related via a swab test, and were therefore not computed in the official statistics.

From an analysis of mortality by age, it may be concluded that deaths from direct or indirect Covid-19 causes severely affected people aged 70 years or older. Given that institutions in Italy do not provide information on the age of deceased at regional level, the present study devised a method for estimating this value and for zeroing in on the repercussions of Covid-19 on different age groups. Our study conclusively shows that one in three people aged 70 or over who died in March 2020 died for causes related to the onset of the Covid-19 epidemic.

References

Bucci, E., et al., 2020. Verso una stima di morti dirette e indirette per Covid. scienzainrete.it. Accessed December 2020 from https://www.scienzainrete.it/articolo/verso-stima-di-morti-dirette-e-indirette-covid/enrico-bucci-luca-leuzzi-enzo-marinari.

Buonanno, P., Galletta, S., Puca, M., 2020. Estimating the severity of Covid-19: evidence from the Italian epicenter. SSRN. accessed December 2020 from https://doi.org/10.2139/ssrn.3567093.

Fanelli, D., Piazza, F., 2020. Analysis and forecast of Covid-19 spreading in China, Italy and France. Chaos, Solitons Fractals 134. https://doi.org/10.1016/j.chaos.2020.109761.

ISTAT, 2020. Decessi per il complesso delle cause. Periodo gennaio-agosto 2020, accessed December 2020 from https://www.istat.it/it/files/2020/03/nota-decessi-22-ottobre2020.pdf.

ISTAT, Istituto Superiore di Sanita, 2020a. Impatto dell'epidemia Covid-19 sulla mortalità totale della popolazione residente. Primo trimestre 2020, accessed December 2020 from https://www.epicentro.iss.it/coronavirus/pdf/Rapporto_Istat_ISS.pdf.

ISTAT, Istituto Superiore di Sanita, 2020b. Impatto dell'epidemia Covid-19 sulla mortalità totale della popolazione residente. Primo quadrimestre. accessed December 2020 from https://www.epicentro.iss.it/coronavirus/pdf/Rapp_Istat_Iss_3Giugno.pdf.

ISTAT, Istituto Superiore di Sanita, 2020c. Impatto dell'epidemia Covid-19 sulla mortalità totale della popolazione residente. Periodo gennaio-maggio. accessed December 2020 from https://www.iss.it/documents/20126/0/Rapp_Istat_Iss_FINALE+2020_.pdf/6f215a7f-98b0-a016-088b-83001190e4a7?t=1609327663819.

Marino, D., Musolino, D., 2020. Differenze regionali nella mortalità ufficiale e "nascosta" da Covid-19: il caso Lombardia nel contesto nazionale e internazionale. economiaepolitica 19 (1). https://www.economiaepolitica.it/l-analisi/rt-differenze-regionali-mortalita-ufficiale-da-covid-19-nord-centro-sud-italia-lombardia/.

Medford, A., Trias-Llimós, S., 2020. Population age structure only partially explains the large number of Covid-19 deaths at the oldest ages. Demogr. Res. 43, 533–544. https://doi.org/10.4054/DemRes.2020.43.19.

Modi, C., et al., 2020. How deadly is Covid-19? A rigorous analysis of excess mortality and age-dependent fatality rates in Italy. medRxiv. https://doi.org/10.1101/2020.04.15.20067074.

Murgante, B., Borruso, G., Balletto, G., Castiglia, P., Dettori, M., 2020. Why Italy first? Health, geographical and planning aspects of the Covid-19 outbreak. Sustainability 12, 5064. https://doi.org/10.3390/su12125064.

Onder, G., Rezza, G., Brusaferro, S., 2020. Case-fatality rate and characteristics of patients dyingin relation to Covid-19 in Italy. JAMA 323 (18), 1775–1776. https://doi.org/10.1001/jama.2020.4683.

Yuan, J., et al., 2020. Monitoring transmissibility and mortality of Covid-19 in Europe. Int. J. Infect. Dis. 95, 311–315.

2.2

Mortality and severity of infection in Lombardy

Ilia Negri and Marcella Mazzoleni

2.2.1 Introduction

Lombardy was the region most affected by Covid-19 in Italy. In this section we show in detail which territories and which population groups in the region were the most affected. This is done by analyzing mortality data in the months of March and April 2020 and by focusing on the outbreak in Lombardy. What emerges are three specific sub-regions, in a certain sense three "Lombardies," ranked according to outbreak severity across this region. It turns out that this breakdown corresponds to three territories with different geographical features such as population, morphology, and economic activity.

Various data sources were used. Daily reports from the ISS (Istituto Superiore di Sanità—Italian Higher Health Institute) account for the number of new Covid-19 deaths for swab-tested subjects who tested positive. However, such data comprise solely the overall number of deaths at the provincial level and the death toll by age group at the national level. ISTAT (the Italian National Institute of Statistics), on the other hand, publishes tables which list all deaths in Italy arranged by municipality and by age in the first 8 months of the year, from 2015 to 2020.

ISTAT (ISTAT and Istituto Superiore di Sanità, 2020a) published an initial report over mortality in Italy for the first quarter of 2020, with a focus on deaths ascribed to Covid-19. This ISTAT report recorded an excess of deaths, by comparing the difference between deaths in 2020 and in previous years (from 2015 to 2019) with the official Covid-19 death toll. A second ISTAT (ISTAT and Istituto Superiore di Sanità, 2020b) report elaborated that this excess could be due to other three separate causes, namely: (1) additional mortality due to Covid-19 (deaths where no swab-tests were performed); (2) indirect mortality associated with Covid-19 (deaths attributable to organ dysfunction, such as heart or kidney, as a consequence of disease triggered by the virus in untested people) and, (3) finally, indirect mortality rate unrelated to the virus but caused by a major strain on the health system and by people's reluctance to be hospitalized in the most affected areas.

ISTAT mortality tables and official mortality rates from ISS and other sources have been discussed at length in the literature, with most studies agreeing on the gravity of the Covid-19 outcome in the region of Lombardy. Research by Modi et al. (2020), Buonanno et al. (2020), Marino and Musolino (2020), Bucci et al. (2020) and Murgante et al. (2020) attest as much.

In our study, we aim to assess mortality rates in the 12 provinces of Lombardy with a focus on deaths by age either directly due to Covid-19 or indirectly attributable to Covid-19. We also intend to compare these rates with those ascribable to causes other Covid-19 and to the official number of deaths from Covid-19. Our aim is therefore to estimate the mortality rate using two official data sources: Covid-19 mortality in all the provinces, as provided by ISS, and overall mortality (either from Covid-19 or otherwise caused), provided by ISTAT. We will also consider the death toll as of January 1, 2019, provided by ISTAT. As in the previous section of our research, we tap mortality data in Italy for the first 8 months of 2020 as provided by ISTAT via a memo (ISTAT, 2020) published in October 2020. Our discussion is laid out as follows. The next section contains information regarding mortality in Lombardy in March and April. Section 2.2.3 provides an estimate of Covid-19 mortality, which is further discussed by age group in the third section. Research results are discussed in the final section.

2.2.2 Analysis of mortality data in Lombardy

We first turn to an analysis of all-cause mortality in March and April 2020. March was the month that recorded the highest number of deaths. The latest ISTAT report (ISTAT, 2020) on mortality in Table 2.5 indicates that 42,718 people died in Lombardy in the period under review, of which 25,560 only in March, compared to an average mortality of 16,680 over the previous 5 years.

Excess mortality for the 2 months of 2020 we took into account is evident, and is even more marked in the province of Bergamo: deaths in Bergamo in 2020 were 7896, of which 6075 in March, against an average mortality rate of 1701 in previous years. While mortality peaked in Bergamo in March, the province of Milan recorded a death toll of 10,978 for the 2 months (of which 5584 in March and 5394 in April) against an average death rate for these 2 months over the previous 5 years equal to 5170. These results match our analysis on the spread of Covid-19 infection in Chapter 1.3.

To compare mortality across provinces, Table 2.6 shows the overall mortality rate for every 10,000 inhabitants of each province in March, April and March and April together in 2020, as well as average mortality over the previous 5 years in the same period.

It will be noted that the mortality rate for some provinces (Bergamo, Cremona and Lodi), while still very high compared to the regional average, plunged dramatically from a value much above the regional average in March 2020 to a lower value in April. In other provinces (Brescia, Pavia, Mantua, and Lecco) mortality rate decreased from March to April, albeit not as drastically, since rates were still above the regional average. The trend in remaining province remained relatively consistent with the rest of Italy. It should be underlined that the province of Milan belongs to the latter group. Although Milan is currently recording a very high number of cases as of November 2020, its mortality rate by population sits on a par with the rest of Italy. Another striking result of our analysis is that, with regard to the 2 months of

TABLE 2.5 Total deaths for all causes in 2020 and average deaths in the previous 5 years over the same period for March, April and March + April; percentage change by province (calculated from ISTAT mortality data).

Total deaths	March			April			March and April		
	2020	15–19	%var.	2020	15–19	%var.	2020	15–19	%var.
Varese	1098	814	35%	1284	732	75%	2382	1547	54%
Como	872	530	65%	912	494	85%	1784	1024	74%
Sondrio	329	184	79%	331	167	98%	660	351	88%
Milan	5584	2701	107%	5394	2469	118%	10,978	5170	112%
Bergamo	6075	902	574%	1821	800	128%	7896	1701	364%
Brescia	4189	1068	292%	2268	935	143%	6457	2002	223%
Pavia	1464	617	137%	1327	546	143%	2791	1163	140%
Cremona	1917	381	403%	862	326	164%	2779	706	294%
Mantua	916	417	120%	737	368	100%	1653	785	111%
Lecco	798	284	181%	620	270	130%	1418	554	156%
Lodi	968	205	372%	350	189	85%	1318	394	235%
Monza and Brianza	1350	677	99%	1252	607	106%	2602	1283	103%
Lombardy	*25,560*	*8780*	*191%*	*17,158*	*7903*	*117%*	*42,718*	*16,680*	*156%*

TABLE 2.6 Mortality rate for all causes for each 10,000 inhabitants in 2020 and average deaths over the previous 5 years in the same period for March, April and March + April (calculated from ISTAT mortality data).

Mortality rate for each 10,000 inhabitants	March		April		March and April	
	2020	2015–2019	2020	2015–2019	2020	2015–2019
Varese	12	9	14	8	27	17
Como	15	9	15	8	30	17
Sondrio	18	10	18	9	36	19
Milan	17	8	17	8	34	16
Bergamo	55	8	16	7	71	15
Brescia	33	8	18	7	51	16
Pavia	27	11	24	10	51	21
Cremona	53	11	24	9	77	20
Mantua	22	10	18	9	40	19
Lecco	24	8	18	8	42	16
Lodi	42	9	15	8	57	17
Monza and Brianza	15	8	14	7	30	15
Lombardy	*25*	*9*	*17*	*8*	*42*	*17*

March and April 2020 in which the Covid-19 outbreak surged, the province with the highest mortality rates was no longer Bergamo but Cremona.

2.2.3 Covid-19 mortality estimate

To reach an accurate assessment of excess mortality in the 2 months of March and April 2020, we estimated the number of deaths from causes attributable to Covid-19, indicated with Covid+, which includes confirmed Covid-19 deaths and an estimate of undiagnosed Covid-19 deaths. These estimates were obtained using the method laid out in Section 2.1.2: namely, each r area indicates a Lombard province. Mortality data by province provided by ISTAT (ISTAT and Istituto Superiore di Sanità, 2020b) were used as constraints. The results are summarized in Table 2.7 and Fig. 2.4.

A comparison of March and April death rates for Covid-19 and Covid+ in Table 2.7 has the province of Bergamo showing the highest difference in absolute number of deaths and the highest mortality rate. In addition, for every 100 official deaths from Covid-19 there are an estimated 207 deaths from Covid+, with a difference of 107 deaths not counted in official data.

With respect to the relationship between Covid+ and Covid-19, the province of Lecco has the highest ratio, since for every 100 deaths from Covid-19 there are 209 deaths from Covid+, with a difference of 109 deaths unaccounted for in official data. Although the first case of Covid-19 was recorded in the province of Lodi, the Covid+ and Covid-19 ratio in that area

TABLE 2.7 Covid-19 and Covid+ mortality rates each for 10,000 inhabitants in March and April and the number of Covid-19 and Covid+ deaths (estimated from ISTAT mortality data).

	Covid+ mortality rate	Covid-19 mortality rate	Covid+ deaths	Covid-19 deaths	Covid+/ covid
Varese	9	4	835	383	2.18
Como	13	8	760	485	1.57
Sondrio	17	10	309	179	1.73
Milan	18	11	5808	3450	1.68
Bergamo	56	27	6195	2994	2.07
Brescia	35	19	4455	2466	1.81
Pavia	30	19	1628	1047	1.55
Cremona	58	29	2073	1038	2.00
Mantua	21	15	868	616	1.41
Lecco	26	12	864	413	2.09
Lodi	40	29	924	658	1.40
Monza and Brianza	15	9	1319	751	1.76
Lombardy	*26*	*14*	*26,038*	*14,480*	*1.80*

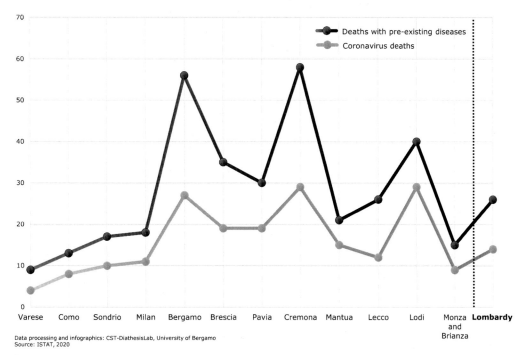

FIG. 2.4   Covid-19 and Covid+ mortality rates in March and April 2020 by province (estimated from ISTAT mortality data).

remained low. This value settles at 1.40, i.e., for every 100 deaths from Covid-19, 140 deaths from Covid+ are estimated with a difference of 40 uncounted deaths. These results are consistent with the analysis on contagion in Lombardy reported in Chapter 2.1.

Fig. 2.5 shows a color-coded map, where different shades of green are used to indicate mortality rates, and variable reliefs obtained by extruding province surfaces mark the absolute number of Covid+ deaths in April and March 2020. The highest mortality rate values are recorded for Bergamo and Cremona, color-coded in dark green. However, the Cremona area on the map appears in lower relief than Bergamo, because the number of deaths in absolute terms in the latter province was lower than in Bergamo. Conversely, other provinces such as Brescia and Milan have very high relief bars but lower mortality rates. Nonetheless, it should be noted that the mortality rate in Brescia is higher than in Milan, color-coded in dark green, although the two provinces have a very similar estimated number of Covid+ deaths (see the last column of Table 2.6). Lecco, Lodi and Pavia show a fairly high mortality rate (darker shades of green) but low extrusion relief, while the other provinces show a low mortality rate (light green color) and a very low relief, markers of diminished severity. Cross-referencing of two different data sets in map form clearly outlines three separate subregions in Lombardy.

Indeed, the three "Lombardies" stand out quite clearly in Fig. 2.5. The provinces of Bergamo, Brescia and Milan make up a first sub-region, in which Covid-19 impact was more severe and mortality rates were high both in absolute terms and by population. These are the three Lombard provinces characterized by dense urbanization and intense productivity. Then there are the provinces of Pavia, Lodi, Cremona, and Mantua, where the impact of

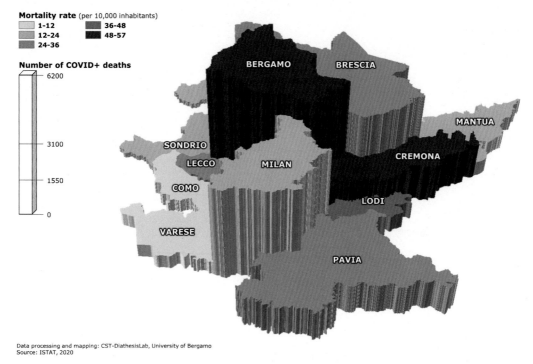

Data processing and mapping: CST-DiathesisLab, University of Bergamo
Source: ISTAT, 2020

FIG. 2.5   Number of Covid+ deaths and mortality rate in March and April 2020.

Covid-19 was less marked if compared with Bergamo, Milan and Brescia, while still higher than in the rest of Italy. This is an area of Lombardy with a strong agriculture-driven productive network, made up of smaller urban areas. Finally, the other provinces of Lombardy are the ones less affected by the first Covid-19 outbreak. Contagion impact for these provinces is similar to the rest of Italy.

Fig. 2.6 is an anamorphic map which uses surface distortion (dilation or contraction) to visualize data based on the number of Covid+ deaths in March and April 2020 for each municipality. Color-coding shows the death rate per 1000 inhabitants for each Lombard municipality.

Reflexive cartography of this kind—which presents localized and cross-referenced data—enables researchers to zero-in on mortality data which impact each municipality while at the same time providing a wider, synoptic model which makes sense of province distribution. This model accurately plots the social impact of Covid-19 mortality and promotes targeted research for the causes of such differences in other socio-territorial factors such as type of settlement, mobility, pollution or other.

While Milan is recording a high value of deaths, given by the dilated surface, it has a low mortality rate value (color-coded in light green). On the contrary, other areas such as the municipalities at the north of Bergamo and around Cremona have a higher mortality rate (color-coded in dark green) but a lower number of deaths, visually signaled by a contraction of its surface. The cities of Bergamo, Brescia and Milan record a high number of deaths, but a discontinuous mortality rate, given by different color-coding. From this map it is evident the division of Lombardy into three areas, but this evidence appears in a different way with

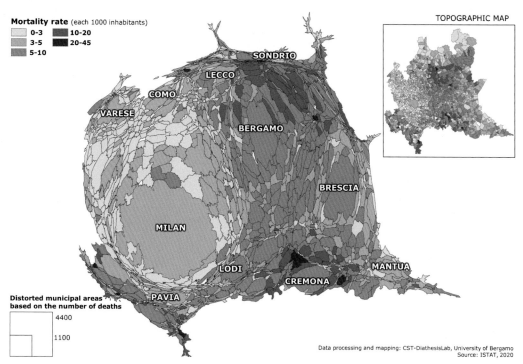

Mortality rate (each 1000 inhabitants)
- 0-3
- 3-5
- 5-10
- 10-20
- 20-45

TOPOGRAPHIC MAP

SONDRIO
LECCO
COMO
VARESE
BERGAMO
BRESCIA
MILAN
LODI
MANTUA
CREMONA
PAVIA

Distorted municipal areas based on the number of deaths
4400
1100

Data processing and mapping: CST-DiathesisLab, University of Bergamo
Source: ISTAT, 2020

FIG. 2.6 Number of deaths and mortality rate for each municipality of Lombardy in March and April 2020.

respect to that in Fig. 2.5. Clearly, Covid-19 affected almost exclusively the provinces of Milan, Bergamo and Brescia, which visually expand to cover nearly the entire surface of Lombardy, followed by the southern provinces (Pavia Lodi, Cremona and Mantua). The provinces to the north-west (Como, Varese and Sondrio) are hardly visible, as they are almost completely crushed by the others: a clear indication of the epidemic's minor impact there.

2.2.4 Analysis of mortality by age

The Covid-19 outbreak affected age groups differently. Oldest age groups are well known to have recorded the highest absolute numbers of deaths.

Cross-referencing of mortality data from all causes shows that subjects over the age of 90 were the most affected in the months of March and April 2020, as shown in Table 2.8, which visualizes all-cause mortality rates in 2020 and 2019 for oldest age groups. With regard to the oldest age group (people over 90 years of age), up against a rate that was around 356 deaths per 10,000 inhabitants in 2019 throughout the whole of Lombardy, for 2020 we witness a mortality rate higher than 1400 deaths per 10,000 inhabitants in the provinces of Bergamo and Cremona. As age decreases, rates do fall. However, set up against 2019 mortality rates, the difference remains higher for the provinces of Bergamo, Cremona, Lodi, Brescia, and Pavia.

Table 2.9 shows how excess mortality recorded in the oldest age groups is almost exclusively to be ascribed to Covid-19 or related causes, since differences between provinces in mortality rate from other causes are negligible.

TABLE 2.8 All-causes mortality rates for each 10,000 inhabitants in March and April 2020 (calculated from ISTAT mortality data).

Mortality rate for 2020 and 2019	50–59		60–69		70–79		80–89		≥90	
	2019	2020	2019	2020	2019	2020	2019	2020	2019	2020
Varese	4	5	11	18	33	46	112	179	344	589
Como	5	7	11	20	39	59	112	208	378	638
Sondrio	4	10	15	27	38	72	113	252	422	628
Milan	5	7	12	21	35	66	105	233	328	751
Bergamo	4	15	12	58	36	193	113	563	361	1429
Brescia	4	10	11	37	35	126	113	381	342	1065
Pavia	7	9	14	38	37	105	129	311	411	845
Cremona	7	12	13	45	35	165	129	493	372	1467
Mantua	5	7	13	20	31	75	112	276	411	723
Lecco	4	8	10	22	33	87	113	293	400	883
Lodi	5	13	10	47	33	172	153	397	391	1103
Monza and Brianza	5	6	11	19	32	62	104	220	348	650
Lombardy	*5*	*8*	*12*	*29*	*35*	*93*	*112*	*297*	*356*	*865*

TABLE 2.9 The mortality rate for each 10,000 inhabitants in the period March and April 2020 divided by causes linked to Covid-19 and other causes (estimated from ISTAT mortality data).

Mortality for covid+ and other causes	70–79		80–89		≥90	
	Covid+	Other causes	Covid+	Other causes	Covid+	Other causes
Varese	13	33	66	112	252	337
Como	23	36	95	113	299	339
Sondrio	32	41	138	114	236	392
Milan	32	34	131	102	444	307
Bergamo	157	36	448	115	1096	333
Brescia	93	33	267	114	717	348
Pavia	64	40	183	128	494	351
Cremona	127	38	367	126	1124	343
Mantua	42	33	155	120	360	363
Lecco	56	31	182	112	553	330
Lodi	133	39	270	127	728	374
Monza and Brianza	31	32	118	102	359	292
Lombardy	*58*	*35*	*186*	*111*	*536*	*329*

As surmised by existing literature, the Covid-19 epidemic did have a significant impact on mortality in oldest ages in Lombardy and in the provinces of Bergamo, Cremona, and Lodi, but also in the provinces of Pavia and Brescia. The impact was more attenuated, but still comparatively high, in the other provinces. As expected, throughout all the provinces, oldest age groups recorded the highest mortality rates, which confirms that the disease strongly affects the elderly.

2.2.5 Conclusions

We considered data provided by ISTAT on total mortality in Italy in the first 8 months of 2020 and zeroed in on the months of March and April 2020, the two critical periods in which Italy had to come to terms with the novel SARS CoV-2 virus threat, responsible for the Covid-19 disease.

Since targeted information on age-related mortality is missing, especially for regions, such as Lombardy, where the impact of Covid-19 was significant, we estimated mortality and mortality rate by integrating data disclosed by the ISS (without indications of age) with those disclosed by ISTAT for the months of March and April 2020. Estimates were calculated for each of the 12 Lombardy provinces starting from mortality data in March and April over previous years from 2015 to 2019, a time frame which was adopted for the purposes of comparison and control.

The results confirm that Covid-19 impact in Italy was significant. More specifically, Lombardy was the region that, in addition to the highest absolute numbers of deaths, recorded the highest rates of variation and the highest mortality, as shown in the previous section of this chapter. A meta-analysis of mortality data in the region of Lombardy suggests that, for the purposes of data interpretation, the region may be divided into three macro-areas. The first macro-area is made up of the provinces of Bergamo, Milan, and Brescia, where both the absolute numbers of deaths and mortality rates were very high, between three and four times those of previous years. The second macro-area consists of their neighboring provinces: Pavia, Lodi, Cremona, and Mantua. Here the impact of mortality was greater than in other regions of Italy, but lower than the first three provinces. The third group comprises the other provinces, where the incidence of mortality was lower and generally comparable to the Italian average.

References

Bucci, E., et al., 2020. Verso una stima di morti dirette e indirette per Covid. scienzainrete.it. accessed December 2020 from https://www.scienzainrete.it/articolo/verso-stima-di-morti-dirette-e-indirette-covid/enrico-bucci-luca-leuzzi-enzo-marinari.

Buonanno, P., Galletta, S., Puca, M., 2020. Estimating the Severity of Covid-19: Evidence from the Italian Epicenter. SSRN. accessed December 2020 from https://doi.org/10.2139/ssrn.3567093.

ISTAT, 2020. Decessi per il complesso delle cause. Periodo gennaio-agosto 2020, accessed December 2020 from https://www.istat.it/it/files/2020/03/nota-decessi-22-ottobre2020.pdf.

ISTAT, Istituto Superiore di Sanità, 2020a. Impatto dell'epidemia Covid-19 sulla mortalità totale della popolazione residente. Primo trimestre 2020, accessed December 2020 from https://www.epicentro.iss.it/coronavirus/pdf/Rapporto_Istat_ISS.pdf.

ISTAT, Istituto Superiore di Sanità, 2020b. Impatto dell'epidemia Covid-19 sulla mortalità totale della popolazione residente. Primo quadrimestre 2020, accessed December 2020 from https://www.epicentro.iss.it/coronavirus/pdf/Rapp_Istat_Iss_3Giugno.pdf.

Marino, D., Musolino, D., 2020. Differenze regionali nella mortalità ufficiale e "nascosta" da Covid-19: il caso Lombardia nel contesto nazionale e internazionale. economiaepolitica 19 (1). https://www.economiaepolitica.it/l-analisi/rt-differenze-regionali-mortalita-ufficiale-da-covid-19-nord-centro-sud-italia-lombardia/.

Modi, C., et al., 2020. How deadly is Covid-19? A rigorous analysis of excess mortality and age-dependent fatality rates in Italy. medRxiv. https://doi.org/10.1101/2020.04.15.20067074.

Murgante, B., Borruso, G., Balletto, G., Castiglia, P., Dettori, M., 2020. Why Italy first? Health, geographical and planning aspects of the Covid-19 outbreak. Sustainability 12 (5064). https://doi.org/10.3390/su12125064.

Chapter 3. Mobility and urbanization

3.1

Commuting in Europe and Italy

Alessandra Ghisalberti

3.1.1 Commuting between proximity and reticularity

Commuting is a mobility factor that may have impacted the speed and intensity of the spread of Covid-19 in spring 2020.[a] As they display the reticular nature of inhabitants' movements, commuting patterns attest to the intense social interaction which characterizes some territories. They also determine crowding in specific time slots, especially on mass transport.[b] As amply noted in the literature,[c] mass transport entails close contact between the inhabitants and turns them into vectors of contagion, which facilitates propagation. Specifically, as it innervates the urbanized clusters of globalization, inhabitants' mobility may have favored contagion both by proximity, or via contact between people who inhabit contiguous places, and by reticularity, that is via the multiple flows of people who commute daily between distant sites, thereby circulating the virus through the use of collective means of transport (Casti and Adobati, 2020).

Along these lines, the present study illustrates commuting practices in Europe, and then focuses on the Italian territory, which saw the emergence of a major Covid-19 epicenter between the months of February and June 2020. In fact, it is precisely in the northern regions of Italy that severe viral outbreaks can be traced back to mass transport commutes, a salient

---

[a] This study embraces the theoretical underpinnings and the methodological approach developed by Emanuela Casti and illustrated in the introduction to this volume. Specifically, Casti emphasizes that the different modes of contagion are attributable to the inherent fragility of contemporary habitation patterns, which are mobile and urbanized. This fragility calls for a radical rethinking of future territorial policies. See also: Casti (2020a,b), especially page 75.

[b] Commuting is a form of mobility that characterizes the contemporary world and makes for ever greater acceleration. Its outcomes are "hyper-mobile" societies based on the endless movement of inhabitants. It involves temporary and recursive changes of place, via movements inhabitants perform for work or study reasons. For an in-depth discussion of globalization, see: Lussault and Stock (2003); Urry (2007).

[c] Among the many studies that focus on the relation between Covid-19 and commuting, the following investigate the relation between public transport and health care and zero in on Italy's case: Laverty et al. (2020); Cartenì et al. (2020).

feature throughout the Po Valley which may have facilitated overcrowding. In this study, this form of mobility is rendered via reflexive cartography which relies on a range of techniques to show commuters distribution by cross-referencing it to demographics.[d] What results is a composite picture of Europe, made up of more densely populated areas heavily affected by commuting, especially across various states of central-western Europe. In some cases—such as Northern Italy—these areas were severely affected by the SARS-CoV-2 epidemic.

3.1.2 The European context of mobility

Commuting is a significant phenomenon in Europe, since it involves over 18 million inhabitants who move for work, equal to approximately 8.3% of its 220 million inhabitants. Commuting has increased in recent years, due to changes in the organization of production systems that have led to greater flexibility in worker mobility, but also due to enhancements in transport and communication infrastructures, which has in turn streamlined the movement of goods and services and has led to an expansion of commuter routes.[e]

Specifically, we may investigate this form of mobility in Europe by analyzing two Eurostat indicators: (1) a *commuter index* calculated on active population members who, for work reasons and at least once a week, move inside or outside European regions, otherwise known as level-2 European Territorial Units for Statistics (NUTS2), which delineate the administrative boundaries of regions; and (2) work-related *mobility areas* characterized by dynamics which include commuting along with other forms of work mobility exceeding 15% of active population members. This indicator identifies areas which surround cities and on which cities are heavily dependent, for reasons related to the labor market. Although such datasets fall short of conveying individual commuter routes and, consequently, also the rhizome-like aspect of commuting, they enable us to pinpoint European regions with the largest number of commuting inhabitants, highlighting their distribution within individual countries and recording their significant presence in metropolitan areas.

With regard to the first indicator,[f] mosaic map (Fig. 3.1) displays the commuters' index among active population members in the European regions by color-coding in various shades of blue, as well as on the basemap anamorphically distorted to reflect the number of residents. By cross-referencing multiple datasets, this map provides an effective visualization of discontinuity across Europe's territory in terms of demographics and commutes. As regards the first indicator, the basemap highlights the most densely populated areas such as, for example,

---

[d] This study adopts the notion of "reflexive cartography" established by Emanuela Casti, to envisage maps that include indices, integrate wider datasets or use anamorphic representation in order to bring out social and territorial dynamics; see: Casti (2015).

[e] We rely on Eurostat data on commuting, understood as infra-regional or inter-regional movement which occurs at least once a week, from places of residence to workplaces, and involves employed individuals aged 15–64 in European countries. See: Eurostat (2019).

[f] As mentioned above, data have been issued by Eurostat and focus on commuter mobility among active population members. They may be accessed at: https://ec.europa.eu/eurostat/databrowser/view/LFST_R_LFE2ECOMM__custom_135338/default/table?lang=en.

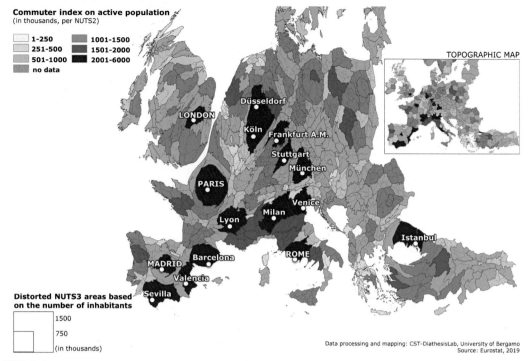

Commuter index on active population (in thousands, per NUTS2)

- 1-250
- 251-500
- 501-1000
- no data
- 1001-1500
- 1501-2000
- 2001-6000

TOPOGRAPHIC MAP

Distorted NUTS3 areas based on the number of inhabitants

1500
750
(in thousands)

Data processing and mapping: CST-DiathesisLab, University of Bergamo
Source: Eurostat, 2019

FIG. 3.1 Commuter index on active population members and anamorphic map of residents in Europe.

the central-northern regions of Italy, the central-southern regions of the United Kingdom or those of the southern coast of Spain. These are matched by various metropolitan clusters such as Paris and Lyon in France, Madrid in Spain or Düsseldorf-Köln and München in Germany. This situation is offset by the less densely populated areas such as, for example, the northern United Kingdom, western France, northern Spain or southern Italy.

If, on the other hand, we wish to consider the commuter index, color-coding in dark blue on the map highlights the European regions where this form of mobility is most frequent. Regions surrounding various capitals may be noticed first, along with major European cities: the areas of London and Paris/Ile-de-France in the central-northern European territory; those of Madrid and Rome, respectively in the western and southern territories of Europe; and finally, the Istanbul area to the east. There are also several areas characterized by a high commuting index around cities which, while not attached political capitals, are densely inhabited and may be highly attractive for work reasons, on account of multiple productive or economic infrastructures. This is the case for the regions of: Düsseldorf, Köln, Frankfurt am Mein, Stuttgart, and München, the beating hearts of German industry; Lombardy, with the Milan conurbation, and the Italian north-east, all the way to Venice, leading regions of the Italian economy; the south-eastern French region of Rhône-Alpe, which extends around Lyon and forms the second major center of economic activity in the country after the capital; finally, the regions of Barcelona, Valencia, and Sevilla in Spain.

Besides recording significant commuter flows and substantial demographics, these areas, as highlighted elsewhere in this volume,[g] also record high levels of air pollution tied to dense urbanization. Also, they feature notable levels of relational intensity between inhabitants and facilitate crowding on mass transport systems in specific time slots. As a result, such areas may have influenced the speed and severity of Covid-19 outbreaks.

Conversely, several less densely populated regions are relatively unaffected by commuting: for example, north-central Spain, western France, central-southern Italy or western Germany. The same applies to some countries in Western and Eastern Europe, such as Portugal, Belgium, Luxembourg or Holland and Poland, Belarus or Ukraine. In absolute terms, these countries were also less affected by contagion in spring 2020.

As we turn to the second indicator, namely *work-related mobility areas*, Fig. 3.2[h] shows that are over 700 such areas of varying extension in Europe.

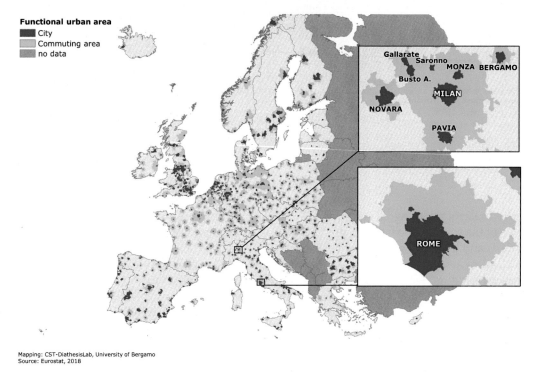

Mapping: CST-DiathesisLab, University of Bergamo
Source: Eurostat, 2018

FIG. 3.2   European cities and areas of work-related mobility, with a detailed cut-out view of Milan and Rome.

[g]See Chapters 1 and 4 in this volume.

[h]The study of urban commuting was conducted by Eurostat, based on research carried out by the OECD (Dijkstra and Poelman, 2012). It refers to cities in the EU member states and the countries of Iceland, Norway, Switzerland and Turkey. It is aimed at identifying *functional urban areas* in Europe, that are areas which feature a densely populated city and a commute zone around it, with a local labor market tightly connected to the city (Eurostat, 2018b).

In some cases, they highlight the centripetal role played by a single urban cluster; in others, they outline a polycentric setup of territory that involves different urban clusters surrounded by the same commuting area.

In the Italian case, we have a dual configuration: Rome belongs to a monocentric work-related mobility area, unlike Milan which presents instead a polycentric kind of mobility layout. The latter, as a political capital, relies on the functional attraction of only on one central core. The second, Italy's "economic capital"[i] is tightly connected with other urban clusters precisely through work-related mobility, which characterize this functional urban area as polycentric. In fact, this commuting area covers an urbanized *continuum* between Milan and the other provincial capitals of Lombardy, such as Monza, Bergamo or Pavia, but also of Piedmont, such as Novara; as well as between Milan and smaller urban centers such as Busto Arsizio, Gallarate or Saronno. This is the main commuting area in Italy, which involves over five million inhabitants in a tight network of inter-regional and inter-provincial connections, where administrative boundaries tend to blur and are relatively insignificant, as amply discussed in Chapter 3.2. The area features a tightly connected territorial fabric which, because of its reliance on mass transport, produces overcrowding. This may well have contributed to the severity of SARS-CoV-2 outbreaks in the northern territory of Italy in the early months of 2020.

3.1.3 A focus on commuting in Italy

Sharper focus in the analysis of commuting in the Italian context, which is the subject of our study, may be achieved by tapping census data published by ISTAT, the Italian Office for National Statistics. Also in this case, we would rely on reflexive cartography for cross-referencing population data with commuting data, with a view to bringing out and emphasizing the social dimension of territory (Casti, 2000, 2015).

Specifically, the map in Fig. 3.3 uses mosaic-like color-coding to visualize a commuter index on active population members and relies on basemap anamorphosis to display areas of Italian municipalities distorted on the basis of the resident population. That makes it possible to zero in on commuting and to note that it affects mostly peri-urban areas. As a matter of fact, clusters such as Milan, Naples or Rome are visually highlighted via anamorphic distortion, which enlarges and foregrounds them on account of their substantial demographics. However, commuting in these metropolitan cities—color-coded in blue—remains limited. Conversely, an area which surrounds urban clusters, color-coded in darker shades of blue, are marked by high levels of commuting.

More specifically, analysis of commuter indices in Italy finds more significant figures in the northern regions, with a definite prevalence of very high levels (between 4001 and 7000) and high (between 2001 and 4000) level of commutes. Conversely, commuting tends to be more restrained—with most average levels between 1001 and 2000, and low levels between 0 and

[i]With regard to the setup of cities in the Italian territory and to Milan's urban reticularity, tied to its multifunction role as the driving force of Italian economy, see: Dematteis (2008).

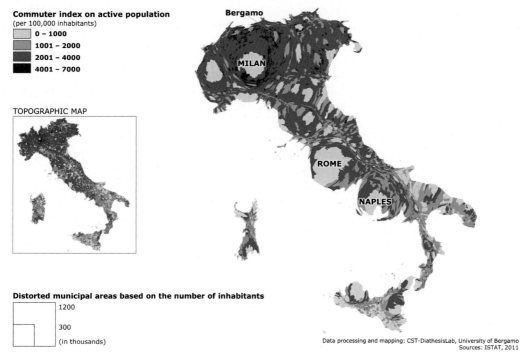

Commuter index on active population
(per 100,000 inhabitants)

☐ 0 – 1000
☐ 1001 – 2000
☐ 2001 – 4000
■ 4001 – 7000

TOPOGRAPHIC MAP

Distorted municipal areas based on the number of inhabitants

1200

300

(in thousands)

Bergamo

MILAN

ROME

NAPLES

Data processing and mapping: CST-DiathesisLab, University of Bergamo
Sources: ISTAT, 2011

FIG. 3.3 Commuter index on active population members and anamorphic map of municipalities based on resident population in Italy.

1000—in the central and, above all, southern areas. This trend makes it possible to envision initial cross-referencing with Covid-19 infection data in Italy and to detect distributive similarities. As discussed in the previous chapters of this volume,[j] outbreaks were considerable in the northern regions of the peninsula, which our analysis shows are mostly involved in commutes, while they tended to be decidedly less extensive in the central-southern regions which record lower commuter numbers.

Also, if we zero in on areas color-coded in dark blue, corresponding to a commuter index on active population members that varies between 4001 and 7000, we find that it involves almost exclusively the conurbation around Milan. We do find a ring around the capital city of Milan, but we also record marked commuting in areas located to the north of the regional capital, especially in the north-eastern area which includes large swathes of the Bergamo province. These are closely-knit areas due to inhabitant mobility, which causes overcrowding on the mass transport network. And these are also areas severely affected by the epidemic, as we underlined earlier.

[j]See especially Chapter 1.2 on the spatial–temporal spread of contagion in Italy. The maps there outlined suggest a tripartite model of Italy's territory that could be named "Three Italies."

3.1.4 Conclusions

Our study has addressed commute data in Europe with a focus on Italy, with the aim to underline how commuting may have affected the severity of Covid-19 outbreaks, since a high-commute region such as Lombardy was the most seriously affected. As the lockdown period showed, however, this type of mobility can be inhibited, via political measures aimed for instance at implementing online technology. This enables residents to carry out work or training activities from home, as well as to organize meetings, manage remote contacts or stagger access to mass-transport. From a political viewpoint, the European Commission is already intervening in the mobility field and has been adopting measures that seek to improve commuting by regulating the transport sector[k] and financing new infrastructure.[l] The Commission also invests in innovation in the mobility sector by promoting technological research.[m] Starting from this political vantage point, we can envisage systematic interventions for managing crowding and contain outbreaks across Europe. Above all, however, we can design new territories to support a novel habitation model based on the enabling potential of smart technologies and on the reduction and staggering of commuter mobility.

[k]With regard to commuter protection, namely via legislation on passenger rights for rail transport (EC-regulation no. 1371/2007 of the European Parliament and Council, dated 23 October 2007, concerning the rights and obligations of rail transport passengers), see: https://eur-lex.europa.eu/legal-content/IT/TXT/?uri=celex:32007R1371).

[l]There exist many projects funded by the European Union to make transport more efficient and sustainable even at a local scale. These include the construction of the second subway line in Warsaw; the introduction of methane-driven buses in Bologna through ERDF (European Regional Development Fund); the promotion of public rail transport against individual car transport; and the reduction of urban pollution, to be achieved by requiring that local authorities purchase new.

[m]Think, for instance, of real-time travel information or of projects for echo-friendly commutes such as the "Bike 2 Work - smart choice for commuters," discussed here: https://ec.europa.eu/energy/intelligent/projects/en/projects/bike2work.

References

Cartenì, A., Di Francesco, L., Martino, M., 2020. How mobility habits influenced the spread of the Covid-19 pandemic: results from the Italian case study. Sci. Total Environ. 741, 1–9. accessed October 2020 from: https://www.sciencedirect.com/science/article/pii/S0048969720340110?via%3Dihub.

Casti, E., 2000. Reality as Representation: The Semiotics of Cartography and the Generation of Meaning. Bergamo University Press, Bergamo.

Casti, E., 2015. Reflexive Cartography: A New Perspective on Mapping. Elsevier, Amsterdam-Waltham.

Casti, E., 2020a. Geografia a 'vele spiegate': analisi territoriale e mapping riflessivo sul Covid-19 in Italia. Doc. Geogr. 1, 61–83.

Casti, E. (Ed.), 2020b. Pourquoi Bergame? Analyser le nombre de testés positifs au Covid-19 à l'aide de la cartographie. De la géolocalisation du phénomène à l'importance de sa dimension territoriale. Anthropocene2050. 30 March, accessed October 2020 from https://medium.com/anthropocene2050/pourquoi-bergame-5b7f1634eede.

Casti, E., Adobati, F. (Eds.), 2020. Mapping riflessivo sul contagio del Covid-19. Dalla localizzazione del fenomeno all'importanza della sua dimensione territoriale. 2° Rapporto di Ricerca, L'evoluzione del contagio in relazione ai territori. CST, Università degli Studi di Bergamo, Bergamo. accessed October 2020 from https://medium.com/cst-diathesislab/2-rapporto-di-ricerca-e566a93431d3.

Dematteis, G. (Ed.), 2008. L'Italia delle città: tra malessere e trasfigurazione, Scenari italiani 2008—Rapporto annuale della Società Geografica Italiana. Società Geografica Italiana, Rome.

Dijkstra, L., Poelman, H., 2012. Cities in Europe the New OECS-EC Definition. OECD, Regional Focus 01. accessed October 2020 from https://ec.europa.eu/regional_policy/sources/docgener/focus/2012_01_city.pdf.

Eurostat, 2018b. Territorial Typologies Manual—Cities, Commuting Zones and Functional Urban Area. European Commission. accessed October 2020 from https://ec.europa.eu/eurostat/statistics-explained/index.php?title=Archive:Statistics_on_commuting_patterns_at_regional_level.

Eurostat, 2019. People on the Move: Statistics on Mobility in Europe. European Commission. accessed October 2020 from https://ec.europa.eu/eurostat/cache/digpub/eumove/.

Laverty, A.A., et al., 2020. Covid-19 presents opportunities and threats to transport and health. J. R. Soc. Med. 113 (7), 251–254. https://doi.org/10.1177/0141076820938997.

Lussault, M., Stock, M., 2003. Mobilité. In: Lévy, J., Lussault, M. (Eds.), Dictionnaire de la Géographie. Editions Belin, Paris, pp. 622–625.

Urry, J., 2007. Mobilities. Poloty Press, Cambridge.

Further reading

Cavallaro, F., Dianin, A., 2019. Cross-border commuting in Central Europe: features, trends and policies. Transp. Policy 78, 86–104.

Ebersold, S., 2020. L'accessibilité, véritable enjeu de société. Le virus de la recherche. PUG.

Eurostat, 2016. Urban Europe—Statistics on Cities, Towns and Suburbs—Working in Cities. Urban Europe: Statistics on Cities, Towns and Suburbs, European Commission. accessed October 2020 from https://ec.europa.eu/eurostat/statistics-explained/pdfscache/50937.pdf.

Eurostat, 2018a. Statistics on Commuting Patterns at Regional Level. European Commission. accessed October 2020 from https://ec.europa.eu/eurostat/statistics-explained/pdfscache/50943.pdf.

Ferro, L., et al. (Eds.), 2018. Moving Cities: Contested Views on Urban Life. Springer, Wiesbaden.

Grieco, M., Urry, J. (Eds.), 2016. Mobilities: New Perspectives on Transport and Society. Routledge, New York.

Hamidi, S., et al., 2020. Does density aggravate the Covid-19 pandemic? J. Am. Plann. Assoc. 86 (4), 495–509. https://doi.org/10.1080/01944363.2020.1777891.

Koehl, A., 2020. Urban transport and Covid-19: challenges and prospects in low- and middle-income countries. Cities Health (Special Issue: Covid-19), 1–6. https://doi.org/10.1080/23748834.2020.1791410. Routledge.

Lapatinas, A., 2020. The Effect of Covid-19 Confinement Policies on Community Mobility Trends in the EU. EUR 30258 EN, Publications Office of the European Union, Luxembourg, https://doi.org/10.2760/875644.

Lévy, J., 2020. L'humanité habite le Covid-19. AOC. Analyse, Opinion, Critique. 26 March 2020, accessed October 2020 from https://aoc.media/analyse/2020/03/25/lhumanite-habite-le-covid-19/.

Lumet, S., Enaudeai, J., 2020. Organisation du territoire européen en temps de Covid-19, entre coopération et repli. Le Grand Continent. accessed October 2020 from https://legrandcontinent.eu/fr/2020/04/01/organisation-du-territoire-europeen-en-temps-de-covid-19-entre-cooperation-et-repli/.

Ralph, D., 2014. Work, Family and Commuting in Europe: The Lives of Euro-Commuters. Palgrave Pivot, London.

3.2

Urbanity and commuting in Lombardy

Emanuela Casti

3.2.1 Urbanity and commuting for reflexive mapping

In spring 2020, the main European epicenter of the Covid-19 epidemic was identified in the territory of Lombardy, a region that features a unique set of socio-territorial phenomena in Italy, namely widespread urbanization, and distinctive commute patterns. In line with our research hypothesis (Casti, 2020a,b), these features are addressed here as factors which de facto impacted the severity of contagion.[a]

Specifically, as regards urbanization, there exists a habitational *continuum*, variously defined by researchers to refer to the virtually seamless extent throughout the Lombard territory, and beyond its borders, reaching as far as the entire Po Valley.[b] Urbanization does not gravitate exclusively around the metropolitan area of Milan. Rather, it is polycentric and multipolar, since it brackets dozens of medium and small towns that shape a complex tangle of settlements. There, widespread and high-density urbanization prevails, within a closely-knit and interdependent functional network. Addressing urbanization from an environmental perspective, the ISPRA Italian National Institute has pointed out that the shape of this network provides the basis for effectively addressing the issue of sustainability and urban resilience. ISPRA has also noted that widespread urbanization not only accelerates loss in landscapes, soils, and related ecosystems, but is also an energy-intensive settlement model which necessitates a substantial framework of both public and private mobility (Marinosci et al., 2015).

[a]Since both may be said to have facilitated crowding and contacts, which propagate contagion by proximity and by reticularity.

[b]The Po Valley is an alluvial plain of the River Po drainage basin: a morphological whole politically divided into the regions of Piedmont, Lombardy, Emilia Romagna, Veneto and Friuli Venezia Giulia. Its urbanization model has been variously defined by analogy to other international concepts (megalopolis, metropolitan region, widespread city, metropolitan archipelago, etc). Such definitions, however, fall short of conveying its distinctive configuration as an urbanized *continuum* notably marked by multidirectional commute patterns. See among others Gottmann (1978); Indovina (2009).

99

Urbanization of this kind is not the result of planning. Rather, it is the legacy of historical stratification originating in the many independent political clusters across the territory, or Signorie (Seignories), which lasted from the Middle Ages to the time of Italian unification, that is, until the second half of the 19th century. As parts of clearly laid out territorial systems, Seignories led to the establishment of a multicentric urban substratum that is culturally distinct yet functionally interdependent.[c] Also, the current urbanization pattern may be defined as pervasive because it has de facto eroded identification boundaries between cities and countryside, which were sharply marked in the past. Urbanization of this kind affects the whole territory not only because it has no spatial continuity, but also because it is marked by the same habitation model conveyed by the notion of *urbanity*.[d] By shifting focus from the material form of built settlements to the way in which individuals inhabit such agglomerations, *urbanity* underlines the mobile character of inhabitants' living based on accessibility systems that make real and virtual connections possible.

This concept postulates that places where individuals give shape to such living patterns need not to be identified solely with cities (although this correlation would seem currently stronger, given that more than half of the world population now lives within urban systems). Rather, these places convey *interconnected living patterns* which are expressed at multiple scales (Djaiz, 2020; Lussault, 2020). It is a habitation pattern engrained in globalization, which unfolds in the intertwining of nodes and connections produced by the dynamic mobility of inhabitants. Inhabitants set out to organize their daily lives not exclusively in the place where they live, but also by interaction with systems that are outside their own, that is the transcalar systems of global networks (Lévy, 2008). In connected living, the common denominator that shapes identity along with the cultural values rooted in the living space, is precisely transcalar mobility. Thanks to globalization, the dynamics induced by the continuous flow of people and information have amplified and accelerated the space–time pace which gives cadence to days and nights, weeks and weekends, the passing of the seasons. On such time-driven cadence, the places in which individuals express their own generalized and transcalar urbanity are metamorphosed.

In its reticular and polycentric setup, the territory of Lombardy expresses this urbanity model. From a material perspective, we are faced with complex urbanization, capable of integrating a set of residential, productive, cultural, and service venues. As regards inhabitants, urbanization entails social dynamics that attach local interests, ways of experiencing

---

[c]For example, the Republic of Venice managed mainland territory via magistrates and offices responsible for its protection, which ensured political unity while preserving local identity; the Duchy of Milan oversaw the control of a territory clearly laid out in a system of courts and parish churches, with the help of ecclesiastical bodies (Casti, 2007).

[d]On urbanity and the fact that it cannot be identified as a lifestyle prerogative of cities but affects the whole territory of complex societies see Lussault (2007, 2017).

mobility and distinctive network connections to a specific place.[e] In short, urbanity enables us to consider the dynamism of inhabitants and the complexity of urbanization as two *sides of the same coin:* a territorial system based on mobility which amplifies contagion in times of pandemic as it facilitates crowding in public spaces. The flow of commuters who, in specific time slots, use collective means of transport attest to the intense social interaction of highly urban areas such as Lombardy.

For this reason, commuting cannot be presented as one among the many forms of mobility that affect urbanization in Lombardy. Nor should it simply be addressed as a mobility type comparable to others in terms of its repercussions on Covid-19 outbreaks. Rather, it is one of the primary causes on which contagion severity depends. Careful reflection is on this issue is called for.

The Wikipedia entry for "commuting" defines it as *periodically recurring travel, daily or weekly, between one's place of residence and place of work, or study, on the part of individuals who usually rely on public transport to move beyond the boundary of their residential community.* This definition highlights three aspects of this mobility model: its timing, the means of transport used and the reason for movement. The implications are that one can predict the days or hours in which commuting occurs and that, since it is an instance of collective mobility, it will affect specific groups of people. As anticipated, in times of pandemic this type of mobility puts inhabitants at risk because within a short time frame—typically the opening and closing hours of schools and businesses or offices—it drives large crowds of people into the confined spaces of mass transport, such as trains, subways and buses. Moreover, transport networks rely on high-crowding public spaces or hubs, such as stations, key interconnections where distancing is just as difficult and contact or interaction between citizens is facilitated.

Evidence of the impact of commutes on Covid-19 transmissions comes from contagion monitoring data in Lombardy, which indicate that severe outbreaks are more likely to occur in medium or small settlements where commute rates are higher than in larger cities. Analysis carried out on the Po Valley outbreaks also suggests that, during the "onset" phase, contagion peaks in smaller towns and, in the absence of containment measures, spreads outward only at a later stage, both by *proximity* across neighboring areas and by *reticularity*, reaching far as people from different areas come into contact.[f]

The mobility and density of inhabitants as a whole are not, therefore, to be classified as factors which facilitate contagion and viral spread, as it has been hastily suggested.

---

[e]With regard to the spatial configuration of the pandemic at its beginning in Europe and precisely in March 2020, Jacques Lévy noted that Covid-19 spreads worldwide via mobility hubs (especially stations, ports, airports) where interactions are numerous, as well as in places where forced contact between bodies is extended (mass transport, cinemas, theaters, concert halls, shops, tourist sites, conference venues, universities, hospitals, etc). Based on data collected in March, he claims that contagion rate in Germany is weaker in large cities and in France it is the *Grand Est* and the *Bourgogne-Franche-Comté* areas which record the highest mortality rates, while *l'Île-de-France* seems to have been relatively spared Lévy (2020a,b). Similarly, in the Po Valley, contagion outbreaks did not occur in the cities, but in peri-urban centers which belonged, however, to the wider polycentric conurbation distinctive of this region. Only at a later stage did contagion spread mainly from hospitals and nursing homes, which became high-infection epidemic hotspots.

[f]As illustrated in Chapter 1 of this volume about the evolution of infection in Lombardy and in Chapter 5 on the genesis and development of contagion outbreaks in the territories of Northern Italy.

What matters is, rather, their combination, which is spatially expressed in specific configurations: the crowding movement causes in specific public spaces or the lack of social distance on public transport during commutes.[g] In short, it is evident that epidemic spread is favored by proximity between people, which is made manifest in specific activities such as commuting. After all, one should not forget that commutes in Lombardy have had and continue to play a crucial role in relation to the region's distinctive setup: a tangle of nodes and connections that unfolds in a dense mesh throughout the entire conurbation.

To be sure, ample evidence exists of the role played by globalization in epidemic spread. One needs just remember, for instance, that even medium-sized cities such as Bergamo are now equipped with a transport *hub*, whereby the virus can reach and circulate worldwide at multiple scales.[h] In such globalized places, which feature high levels of relational intensity due to the simultaneous presence of many people,[i] interactions between individuals are considerably more numerous than elsewhere. They also occur in environments that virtually entail a cancellation of social distancing. It should be clear that the present analysis does not aim to examine the diffusion dynamics of contagion. Rather, it seeks possible causes for the severity of outbreaks by addressing socio-territorial aspects of the places where contagion occurs. Along these lines, commuting may be taken as the first major offender.

The latter type of mobility arguably marks the fragility of urban living exacerbated by the Covid-19 pandemic. Even though knowledge of the SARS-CoV-2 virus remains limited, epidemiologists agree that it is largely air-borne. Unsurprisingly, then, close contact with other people inside enclosed spaces albeit for a short time (as on public transport) increases the risk of contagion. This signals one possible path for exploring territorial policies capable of coping with similar pandemics and for rethinking our ways of living.

Along these lines, our analysis of Lombardy taps commuting and population distribution data from a number of sources[j] to shed light on the "urbanity" phenomenon throughout the territory and to process data-driven and algorithm-based maps in the perspective of

---

[g]That commuting should be one of the causes of viral spread is regrettable, since commutes can be scheduled. Arrangements to prevent overcrowding at certain times or days in collective vehicles can be put in place, as in fact has happened in Italy. In Italy, public transport, along with healthcare, are sectors that fall under both national and regional regulations, which makes it harder to effectively implement viral containment measures. On this issue, see Chapter 6 of this volume.

[h]The "Caravaggio" international airport in Orio al Serio-Bergamo is the third in Italy by passenger and goods volume and is a major hub for the Irish low-cost airline Ryanair. It links Italian cities to over a hundred European destinations and to the African coast.

[i]These are places Michel Lussault has defined as "hyper-places," an expression of contemporary mobile living, in which social contacts are perfunctory and interactions limited to sites that share the same purpose. In times of pandemic hyper-places predictably turn into high-risk areas. See Lussault (2017).

[j]These data were issued: (i) by ISTAT (The Italian Office for National Statistics) as part of the 2011 census in order to trace the number of incoming or outgoing commuters for each Lombard municipality; (ii) by the Lombardy Region, via research conducted in 2014 to assess work-related commutes. With regard to the theoretical-methodological framework for issuing cartographic maps capable of bringing out the social dynamics which innervate territory, we will rely on Emanuela Casti's model, as outlined in Casti (2000, 2015).

reflexivity.[k] By cross-referencing data via an anamorphic model, these maps bring out possible ties between commuting and urbanity, thus enabling researchers to probe sets of territorial dynamics that may have had an impact on outbreak severity.

3.2.2 Monitoring commutes in Lombardy

Various types of data published by the Italian Office for National Statistics (ISTAT) and by the Lombardy Region were used to assess commuting in Lombardy. These were processed statistically and algorithmically first, and then translated graphically on the basis of multiple mapping systems. This was necessary both to render the space–time dimension that commuting entails, and to trace the distinctive trajectories commuting follows in Lombardy.

Initially, data produced by ISTAT were cross-referenced to active population data, i.e. data about people between the ages of 15 and 65. This was done to obtain a commuting index which, applied across the region, could pinpoint the areas most impacted by commutes. Subsequently, attention was paid to a *relational intensity index* which charts the prominence of commuting vis a vis other types of mobility. Finally, databases of the Lombardy region were accessed to tap various surveys (based on interviews and statistical data) which show the rhizome-like pattern of commutes in this region.[1] The basic canvas for such data processing is urbanization, within which commuting takes shape. This was obtained from resident population distribution data visualized anamorphically on the basemap. The number of residents expands or contracts the surface areas of Municipalities, indicating housing density and, as a consequence, also the type of urbanization involved (metropolis, metropolitan municipalities, peri-urban area, widespread urbanization, medium or small cities, villages). Flow *reticularity* is thereby given a spatial frame.

Specifically, Fig. 3.4 uses color-coding in varying shades of blue to visualize a commuting index measured on active population by the Municipality. This index is higher (mostly >51% and, in many cases, even >71%) in the Milan metropolitan area—excluding the municipality of Milan—and in the northern area which reaches the territories of Varese and Como; in the Middle Eastern belt that reaches Bergamo from Lodi, defined as the "backbone of contagion" due to the infection intensity recorded and the fact that the first outbreaks were recorded there[m]; finally, in the easternmost offshoot of the plain in the provinces of Cremona and

---

[k]Data processing was part of the Urban Nexus Excellence Initiative Project: *Intelligent modeling and big data mapping for assessing connectivity and regeneration in some European cities*, promoted between 2016 and 2019 by the University of Bergamo with the collaboration of the EPFL-Ecole Polytechnique Fédérale de Lausanne and Anglia Ruskin University, Cambridge.

[1]Commuting data, collected via the 2011 census and published by ISTAT, may be found at the following links, which let users view or download flow data for each Italian province: https://www.istat.it/pendolarismo/grafici_province_cartografia_2011.html; and for each Italian municipality: http://gisportal.istat.it/bt.flussi/. Population data go back to 2011 and concern ISTAT-implemented updates to the post-census demographic balance: http://demo.istat.it/index_e.php; finally, the database on the Origin/Destination matrix processed by the Lombardy Region may be found at: https://www.dati.lombardia.it/Mobilit-e-trasporti/Matrice-OD2014/wbii-r5a6.

[m]See Chapter 1.3 on the contagion in Lombardy in this volume.

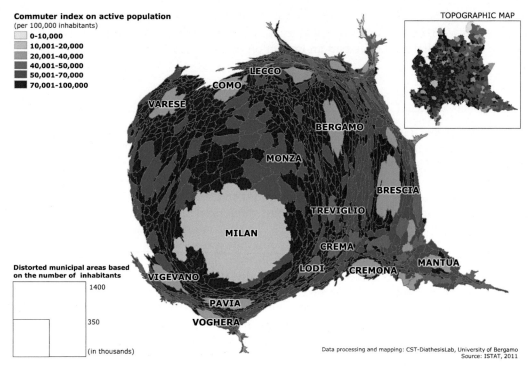

FIG. 3.4   Index of commuter flows in Lombardy compared to the resident population.

Mantua, where data is more varied and includes all color shades. Overall, however, it should be noted that, if we exclude cities, the most widespread color is dark, which confirms that commuting triggers actual territorial osmosis, embracing the entire Lombard conurbation.

With regard to cities, on the other hand, commuting appears to be more contained (Milan has an index of <20%; Varese, Como, Bergamo, Brescia, and Pavia between 20% and 40%). It will be noted however that this index does not record a mobility decrease in the broad sense, but rather informs that mobility within the municipality is greater than commuting. In order to delve deeper, we need to size up the commuting index to other indicators of infra-municipal mobility. We are helped in this by an ISTAT developed indicator, namely the *relational intensity index*.[n] It is an indicator measured on groups of municipalities defined as Local Labor Systems (SLL), which statistically compares commute and infra-municipal flows, making it possible to synopsize overall mobility and thereby derive an "inter-municipal relational level".[o] Fig. 3.5 deploys this index for each SLL in Lombardy, using gray

---

[n]The index was established by ISTAT to compare commute flows for Municipalities that belong to a Local Labor System (i.e., the union of several Municipalities based on active population commute flows) with intra-municipal mobility. By measuring such ratio, it aims to pinpoint Municipalities that heavily rely on commutes. See: www.istat.it/it/informazioni-territoriali-e-cartography/local-work-systems/quality-indicators% C3% A0-sll (ISTAT, 2020).

[o]Readings vary between 0 and 100 and indicate the degree of inter-municipal mobility.

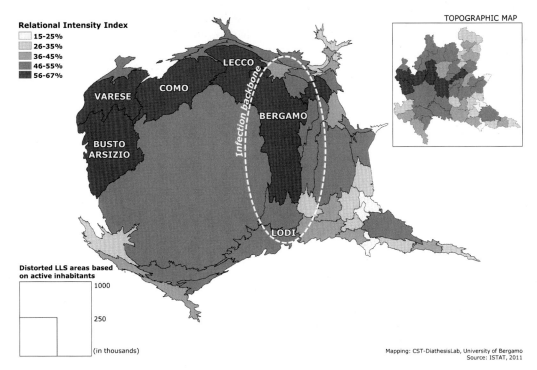

Relational Intensity Index
15-25%
26-35%
36-45%
46-55%
56-67%

TOPOGRAPHIC MAP

LECCO

COMO

VARESE

Infection backbone

BERGAMO

BUSTO ARSIZIO

LODI

Distorted LLS areas based on active inhabitants
1000
250
(in thousands)

Mapping: CST-DiathesisLab, University of Bergamo
Source: ISTAT, 2011

FIG. 3.5   Relational intensity of Local Labor Systems in Lombardy relative to active population.

color-coding to mark size. A vast dark band runs across the map and shows the highest index—with values between 55% and 67%. These include the north of the metropolitan area, characterized by manufacturing industries, which runs from Busto Arsizio to Varese and to the east, passing through Como and Lecco, and reaches Bergamo to continue north-east. The eastern portion of the region, on the other hand, is more variegated and records lower readings—between 46% and 55%. That is tied to a mixed production system strongly aimed at agriculture, which requires a more local or settled labor force. In short, as we already noted in Fig. 3.4, if we exclude the vast northern belt, we discern a region-wide division marked by what we have defined as the "infection backbone" and characterized on average by a high relational intensity index.[P]

Our initial research question was whether there may be links between commuting, urbanity, and the spread of Covid-19. Already at this intermediate stage of analysis, we would claim that the complex articulation of people's movement patterns in Lombardy yields important clues for postulating the existence of territorial factors which favor contagion. The speed, intensity, and severity of outbreaks in either the first or second wave of infection confirm that structural territorial factors—pollution, mobility, urbanization—contribute to viral spread and that commuting plays a crucial role in Lombardy, due to the features we discussed above.

[P] According to ISTAT, relational intensity indices may be related to Covid-19 mortality levels because they impact viral spread (ISTAT, 2020, pp. 86–88).

Accordingly, Lombardy may be defined as the most infected region not so much because it has the largest metropolitan area in the country, Milan, which exerts centripetal attraction. Rather because, despite this presence, there exists a network of marked relational intensity between medium-small cities, which takes the shape of a polycentric urbanization model based on rhizome-like patterns of commuting.

3.2.3 The rhizome-like form of commuting in Lombardy

A resource for monitoring work commutes in Lombardy was established by the Lombardy Region in 2014.[q] It is a database built by integrating the outcome of a study conducted by the "Infrastructure and Mobility" section of the Directorate-General with 2011 data published by ISTAT, which created a matrix on the distance traveled and the number of commuters estimated on an average weekday between February and August. This database was used by CST-DiathesisLab in Bergamo to process the following cartography. Data processing relied both on heuristics, i.e. nonrigorous procedures that enable researchers to predict results to be validated by further analysis, and on an algorithmic model, based on calculations that combine mathematical and probabilistic models.[r]

The first map, in Fig. 3.6, is a heuristic map. Once restructured, data were displayed using lines of varying thickness to trace commutes routes, relative to the number of work commuters; each line was assigned a different color based on the distance traveled in the commute: yellow to indicate short-distance mobility (<20 km); orange to mark medium distance routes (up to 35 km); red for long distance routes. Overall, it emerges that short-range mobility (yellow) concerns the provinces of Monza and Brianza and Varese and partially the provinces of Como, Lecco, Bergamo, Lodi, and Pavia; medium-distance mobility (orange) is prevalent in the provinces of Cremona and Mantua; finally, long-distance mobility (red) stands out in the rest of the region. Color-coding shades, albeit superimposed, bring out a "ribbon" of short-distance commutes that affects a large part of the region north of Milan to which Varese, Como, Monza, Lecco, and Bergamo belong. This tapers off, however, as we move south or turn to the city of Milan, for the reasons we have already had a chance to examine. The most marginal areas—that is the mountainous part of the provinces of Bergamo, Lecco, and Como—and part of the province of Brescia show a reticular layout of medium-long distance commutes, with flows of more than 20 km, as for the province of Sondrio; in the south-eastern area of the region, however, in the provinces of Cremona and Mantua,

[q]For further details, see the Lombardy Region website links which illustrate the method used for building the database via a combination of transport modeling, online questionnaires, face-to-face interviews, analysis of available surveys and of existing demand: https://www.regione.lombardia.it/wps/portal/istituzionale/ HP/DettaglioServizio/servizi-e-informazioni/imprese/imprese-di-trasporto-e-logistica/ser-matrice-od-infr.

[r]Data assembled by the Lombardy Region were restructured in an intelligible format via a Geocoding procedure which georeferenced all the locations in the matrix and calculated the actual number of work commuters, for each commute leg, as estimated by the Region. The resulting maps were assembled at the CST-Diathesis Lab and processed by computer programmer Daniele Ciriello, as part of the Urban Nexus Excellence Initiative project, funded by the University of Bergamo.

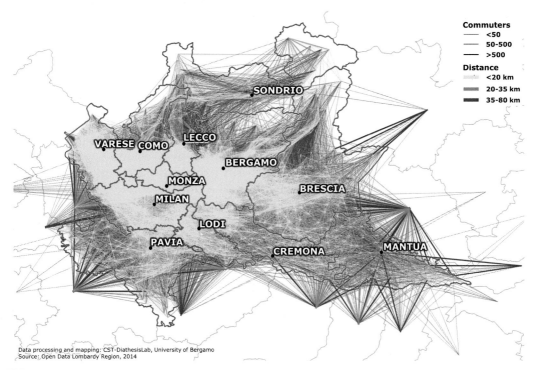

FIG. 3.6 Big data and work commutes in Lombardy: a rhizome-like movement.

commuting occurs at various distances. Ultimately, data processing replicates and confirms what was observed in the previous Fig. 3.5, which attests to the validity of the heuristics.

If, however, the same data is processed using the COMBO algorithm,[s] for automatic detection of "communities" (understood as a dense network of nodes that builds up a closely-knit agglomeration) and if these communities are spatialized and matched to the respective Lombard territories, the map brings out a new dimension of work commuting: an individual dimension which opens up new paths of reflection on the rhizome-like spatiality of individuals grouped together on the basis of the same commuting behavior.[t]

It seems appropriate, at this point, to recall the key notion of a rhizome-like spatiality, whereby contemporary cities are said to take on a polycentric and reticular configuration (Lévy et al., 2016). We may recall that the rhizome, metaphorically approached by two French philosophers (Deleuze and Guattari, 1980), has been adopted by the social sciences to pinpoint network sets that are characterized by the absence of topographical boundaries

[s]This is an open-source algorithm developed by the SENSEable City Lab of the Massachusetts Institute of Technology (MIT): http://senseable.mit.edu/.

[t]It should be noted that the COMBO algorithm associates flows between each node regardless of geolocation, since algorithms of this type are used in social media to identify connections between different entities and visualize them in the form of communities, bringing out networks made up of nodes that are marked by strong interaction.

(Lévy, 2003; Regnauld, 2012): rhizome-like space is the transcalar space of individual action. The reticular pattern drawn by a rhizome is based on a network that has neither beginning nor end, and lacks clearly outlined boundaries: it is the outcome of an individual's isolated experience. In the realm of cartography, rhizomes make it possible to recover a crucial chorographic dimension, which effectively brings out social features while avoiding modeling that relies on the mere distribution, juxtaposition, and localization of phenomena (Casti, 2015). This is why the kind of reticularity which recalls the concept of *rhizome* is a multidirectional reticularity, which operates in a multi-localized fashion and firmly resists permanent rooting. The rhizome is the core element of a reticular space, whose dynamic perspective serves to envision the relational space of an individual who lives inside a network.

Fig. 3.7 shows two cartographic views of Lombardy's rhizome-like commute patterns via algorithmic processing of the same database: in Fig. 3.7A differently color-coded communities agglutinate at spatialized points; in Fig. 3.7B vice versa, points are replaced by connecting lines along the origin–destination path of an individual who belongs to a "community" of individuals plotting the same path and identified by the same color. Both maps were also subjected to topographic layout and administrative subdivision constraints, in order to highlight discrepancy between the rhizome-like dimension and the administrative topographic approach. This also emphasizes the muti-directional quality of movement, accordingly defined as rhizome-like (Casti, 2019).

From an interpretative viewpoint, while the topographical rendering of map (a) privileges localization of Municipalities as nodes and thus remains within the limits of a localized model

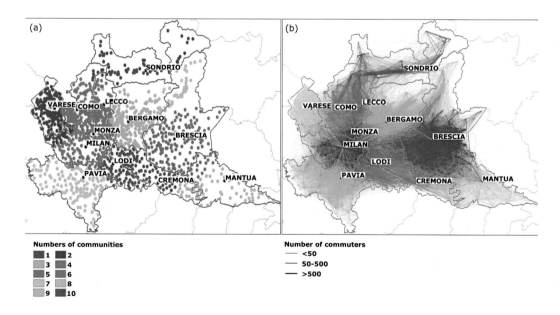

Data processing and mapping: CST-DiathesisLab, University of Bergamo
Source: Open Data Lombardy Region, 2014

FIG. 3.7   Work Commutes: (A) the nodes of new "communities"; (B) "Community" or "Rhizomes."

of commuting based on movement origin, map (b) shows an expansion or contraction of areas which outline new labor territories in relation to commutes. In short, in highlighting the relation between territories based on inhabitant mobility, these representations recall the relational space of individuals who live inside a *network*, that is, individuals who embrace mobility as a constitutive feature of their living-in-places, thereby acquiring the status of *city-users* and inhabitants. More specifically, while the first representation underlines the import of commuter residents who share similar features, the second brings out the multiple directions and journeys that such commuters make. This rhizome-like reticularity is a distinctive feature of Lombardy because, while affected by the attractiveness exerted by Milan, this territory is at the mercy of other centripetal forces and of the attraction exercised by many other medium and small urban centers which make up the connective tissue of the region. The self-consistent whole of the rhizome is well suited to explain the dynamism of the Lombard context. At the same time, it opens up new perspectives to consider "urbanity" as no longer hinged exclusively on the density or diversity criteria of its inhabitants,[u] but as closely related to commuting and to the reticular pattern it produces.

3.2.4 Conclusions

Analysis carried out so far shows that urbanity and commuting in Lombardy are distinctive territorial features in the Italian scenario. Urbanity is spatially expressed via a polycentric urban setup, featuring a major city which, however, does not centralize movement but adds its centripetal force to a complex and variegated mobility pattern, which impacted the speed and intensity Covid-19 contagion. The inhabitants' osmotic dynamics triggers crowding on the public transport networks and causes proximity contacts in public spaces. With regard to contagion, commuting does seem the most implicated type of movement, since it concentrates large numbers of passengers at specific time slots.

The many cartographic representations of this phenomenon we produced via reflexive mapping leave no doubt as to the conclusion that commuting is an extensive phenomenon, but is mostly centered in and around peri-urban areas. Its rhizome-like pattern increases data complexity. However, graphics-driven visualization and algorithmic processing help us grasp the distinctive spread of the Covid-19 epidemic in Lombardy.

Social distancing measures taken in the first wave in the form of a lockdown period; the closure of school activities and non-essential work activities that decreased infection numbers attests to the viability of this interpretative path. Covid-19 has raised awareness of the need to devise a new way of living; a way that may rely on information and communication technology to reduce commute needs and envision living places that are less dependent on collective public transport.

---

[u] And that is not all, for urbanity is also defined as a "caractère proprement urbain d'un espace [...] [here] procède du couplage de la densité et de la diversité des objets de société dans d'espace" (Lussault, 2003, p. 966).

References

Casti, E., 2000. Reality as Representation. The Semiotics of Cartography and the Generation of Meaning. Bergamo University Press, Sestante, Bergamo.

Casti, E., 2007. State, cartography, and territory in the venetian and Lombard renaissance. In: Woodward, D. (Ed.), The History of Cartography. Cartography in the European Renaissance, vol. 3. The University of Chicago Press, Chicago, pp. 874–908.

Casti, E., 2015. Reflexive Cartography. A New Perspective on Mapping. Elsevier, Amsterdam-Waltham.

Casti, E., 2019. Corografia vs topografia (Introduzione). In: Casti, E. (Ed.), La geografia a Bergamo. Nuove sfide per l'analisi territoriale e il mapping. A.Ge.I., Roma, pp. XIII–XXV.

Casti, E., 2020a. Geografia a 'vele spiegate': analisi territoriale e mapping riflessivo sul Covid-19 in Italia. Geogr. Doc. 9, 21–26.

Casti, E. (Ed.), 2020b. Pourquoi Bergame? Analyzer le nombre de testés positifs au Covid-19 à l'Aide de la cartographie. De la geolocalisation du phénomène is the importance of a territorial dimension. Anthropocene2050. 30 March, accessed December 2020 from https://medium.com/anthropocene2050/pourquoi-bergame-5b7f1634eede.

Deleuze, G., Guattari, F., 1980. Mille plateaux. Les Éditions de Minuit, Paris.

Djaiz, D., 2020. La mondialisation malade des ses crisis. Le Grand Continent. https://legrandcontinent.eu/fr/observatoire-coronavirus/. (Accessed 24 August 2020).

Gottmann, J., 1978. Verso una megalopoli della Pianura Padana. In: Muscarà, C. (Ed.), Megalopoli mediterranea. Franco Angeli, Milano, pp. 19–31.

Indovina, F., 2009. Dalla città diffusa all'arcipelago metropolitano. FrancoAngeli, Milano.

ISTAT, 2020. Rapporto annuale 2020. The Situation of the Country. Istituto Nazionale di Statistica.

Lévy, J., 2003. Rhizome. In: Lévy, J., Lussault, M. (Eds.), Dictionnaire de la géographie et de l'espace des sociétés. Editions Belin, Paris, p. 804.

Lévy, J. (Ed.), 2008. L'invention du monde. Une géographie de la mondialisation. Les Presses de Sciences Po, Paris.

Lévy, J., et al., 2016. Rebattre les cartes. Topographie et topologie dans la cartographie contemporaine. Réseaux 195 (1), 17–52.

Lévy, J., 2020a. Rhizome. In: Lévy, J., Lussault, M. (Eds.), Dictionnaire de la géographie et de l'Espace des sociétés. Belin, Paris, p. 804.

Lévy, J., 2020b. L'humanité habite le Covid-19. AOC. Analyse, Opinion, Critique. accessed October 2020 from https://aoc.media/analyse/2020/03/25/lhumanite-habite-le-covid-19/.

Lussault, M., 2003. In: Lévy, J., Lussault, M. (Eds.), Dictionnaire de la géographie et de l'espace des sociétés. Editions Belin, Paris, pp. 966–967.

Lussault, M., 2007. L'homme spatial. La construction sociale de l'espace humain. Seuil, Paris.

Lussault, M., 2017. Hyper-lieux. Les nouvelles géographies de la mondialisation. Seuil, Paris.

Lussault, M., 2020. Le Monde du virus – retourn sur l'eprouve de confinement. AOC. Analyse, Opinion, Critique. https://aoc.media/analyse/2020/05/10/le-monde-du-virus-retour-sur-lepreuve-du-confinement/. (Accessed 24 August 2020).

Marinosci, I., et al., 2015. Forme di urbanizzazione e tipologia insediativa. In: ISPRA, Qualità dell'ambiente urbano—XI Rapporto, pp. 156–173. 2015. accessed December 2020 from https://core.ac.uk/download/pdf/54529419.pdf.

Regnauld, H., 2012. Les concepts de Félix Guattari et Gilles Deleuze et d'espace des géographes. Chimères 1 (76), 3–23.

Further reading

Casti, E., Adobati, F. (Eds.), 2020. Reflective Mapping on the Covid-19 Contagion. Dalla Localizzazione del Fenomeno all'importanza Della Sua Dimensione Territoriale. 2nd Research Report, the Evolution of the Infection in Relation to the Territories. CST, University of Bergamo, Bergamo. accessed December 2020 from https://medium.com/cst-diathesislab/2-research-report-e566a93431d3.

Chatterjee, K., et al., 2020. Commuting and wellbeing: a critical overview of the literature with implications for policy and future research. Transp. Rev. 40 (1), 5–34. https://doi.org/10.1080/08959420.2020.1750543.

Ebersold, S., 2020. Accessibility, veritable enjeu de société. The viruses de la recherche. PUG.

Nelson, D., Rae, A., 2016. An economic geography of the United States: from commutes to megaregions. PLoS One 11 (11), 1–23. 2016 https://doi.org/10.1371/journal.pone.0166083.

Chapter 4. Pollution and territorial diffusion of contagion

4.1

Correlation between atmospheric pollution and contagion intensity in Italy and Lombardy[☆]

Fulvio Adobati and Andrea Azzini

4.1.1 Introduction

Analysis of air pollution levels aims to verify potential correlations between the territorial spread of major atmospheric pollutants, fine dust and nitrogen dioxide and the severity of territorial Covid-19 contagion. A diachronic reading of pollution levels also makes it possible to correlate the impact of restriction measures for combating the SARS-CoV-2 epidemic with data on atmospheric pollution.

[☆]NO$_2$: "Nitrogen dioxide (NO$_2$) is a reddish-brown gas, not very water soluble and toxic. It has a pungent odor and a strong irritating power. The main emission source for nitrogen oxides (NOx = NO + NO$_2$) is vehicular traffic; other sources are civil and industrial heating systems, power plants and a broad range of industrial processes. Nitrogen dioxide is a widespread pollutant that has negative effects on human health. Together with nitrogen monoxide, it contributes to photochemical smog" (Ministry of Health, https://www.salute.gov.it).

PM$_{2.5}$: "The air contains suspended dust which may either be harmless, if of natural origin and present in small quantities, or harmful, if abundant and inhalable. Sources may be of natural or anthropogenic origin (e.g., soot, combustion processes, natural sources and more). Composition varies accordingly. The two main classes of particulate matter are coarse particulate and fine particulate. Coarse particulate matter is made up of particles, including pollen and spores, with a diameter greater than 10 μm (microns). They are generally retained by the upper respiratory tract (nose, larynx). Dust particles with an aerodynamic diameter of less than 10 μm (PM$_{10}$), capable of penetrating the upper respiratory tract (nose, pharynx and trachea), are defined as fine dust. Particles with a diameter of less than 2.5 μm (PM$_{2.5}$) are defined as fine particulate, capable of penetrating deeply into the lungs, especially when breathing occurs through the mouth. For even smaller sizes (ultra-fine particulate matter, UFP or UP) we speak of breathable dust, that is, dust that penetrates deeply into the lungs to the alveoli" (Ministry of Health, https://www.salute.gov.it/).

It needs to be stressed from the start that any estimate of potential correlations should never be taken as final or absolute: atmospheric pollutants do present variations in dispersion and intensity that depend on atmospheric conditions. These considerably affect pollutant accumulation or dispersion, and consequently also monitoring-station measurements. In fact, data reliability increases whenever the reference timeframe is extended, as in the case of monthly or annual averages.

Nonetheless, in atmospheric pollution analysis, space is as important as time, for at least two reasons: (i) analyses conducted in the Lombardy region rely on an IT model that returns approximate information from the data acquired via the network of survey-monitoring stations; (ii) data are mapped with reference to municipal administrative units. In mountain areas, data record a sharp difference in air condition between the urbanized valley bottoms and the sparsely settled area of the medium-high sections on mountain slopes.

It follows that the maps are more reliable in detecting the atmospheric conditions of fields and extended territorial areas, rather than individual areas on a municipal basis.

4.1.2 Links between pollution and contagion

Literature on the causal link between pollution and contagion is vast (Cori and Bianchi, 2020; Setti et al., 2020a,b; Yao et al., 2020) and moves along two major hypotheses.

The first addresses pollution as a viral diffuser, attributing to fine particles (PM_{10} and $PM_{2.5}$) both the role of suspended substrate and the role of viral vector via a "carrier effect"[a] (Setti et al., 2020a). This is based on the propagation dynamics of contagion, which have been found to be tied to high pollution levels in the case of Lombardy, a region severely affected by Covid-19 in the months of March and April 2020. This state-of-the-art hypothesis calls for in-depth study to verify or falsify its assumptions and remains a research topic of significant import.

The second hypothesis is based instead on atmospheric pollution as a key component of territorial health, i.e., as a background environmental condition that affects the health of exposed populations. Territorial fragility is thus ultimately tied to inhabitants' exposure to pollutants (in our case, atmospheric pollutants; Coker et al., 2020; Cori and Bianchi, 2020).

In particular, several studies assume $PM_{2.5}$ atmospheric particulate matter as the main pollutant, specifically associating dispersion of atmospheric particulate with the viral severity.[b]

---

[a] "Our findings suggest that the acceleration of the growth rate observed in Milan could be attributed to a 'boost effect' (a kind of exceptional 'super-spread event') on the viral infectivity of COVID-19, corresponding to the peaks of PM. According to this hypothesis, PM could then act as a potential carrier 'for droplet nuclei', triggering a boost effect on the spread of the virus" (Setti et al., 2020a, p. 7).

[b] "Long-term exposure to ambient air pollutant concentrations is known to cause chronic lung inflammation, a condition that may promote increased severity of Covid-19 syndrome caused by the novel coronavirus (SARS-CoV-2). Our epidemiological analysis uses geographical information (e.g., municipalities) and negative binomial regression to assess whether both ambient $PM_{2.5}$ concentration and excess mortality have a similar spatial distribution" (Coler et al., 2020, p. 1).

If nitrogen dioxide and atmospheric particulate matter pollutants are taken as relevant contagion factors, it is important to look at possible emission sources. As far as PM_{10} is concerned, the main source of emission is domestic heating (namely in the use of woody biomass). Transport-related emissions are also of considerable importance since they determine nearly 20% of PM_{10} emissions. More than 50% of NOx emissions are due to road transport, and this percentage increases in dense urban areas. This condition affects concentration levels of NO_2 in the air, which may be seen as precursors to the diffusion and concentration of atmospheric particulates (PM_{10} and $PM_{2.5}$) as a whole (ISPRA, 2018). Many images show how the Po Valley is among the continental regions with the highest concentration of nitrogen dioxide, as also amply documented in the analytical reports conducted by the European Environment Agency (EEA, 2019).

The image provided by ESA (Fig. 4.1) highlights the concentration of nitrogen dioxide in various European metropolitan areas, and the Po Valley by extension and intensity ranks among highest before the Netherlands.

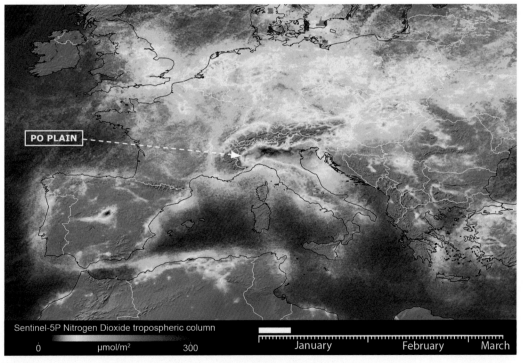

FIG. 4.1 Europe: distribution of nitrogen dioxide pollution, on average values recorded in January 2020. *Source: Modified Copernicus Sentinel data, 2020, processed by ESA, CC BY-SA 3.0 IGO.*

Satellite images provided by ESA (Fig. 4.2) also highlight the differences found in the periods of March–April 2019 and 2020. The effect of restrictions is evident. In the case of NO_2, mobility limitation and vehicular traffic reduction delineate a pollution map that is similar to the standard one but is markedly "lightened" in terms of pollutant concentration, with a substantial halving of pollution levels.

It should be noted that, beyond the current contingency tied to containment measures around the SARS-CoV-2 emergency, a trend towards pollution reduction has been in place

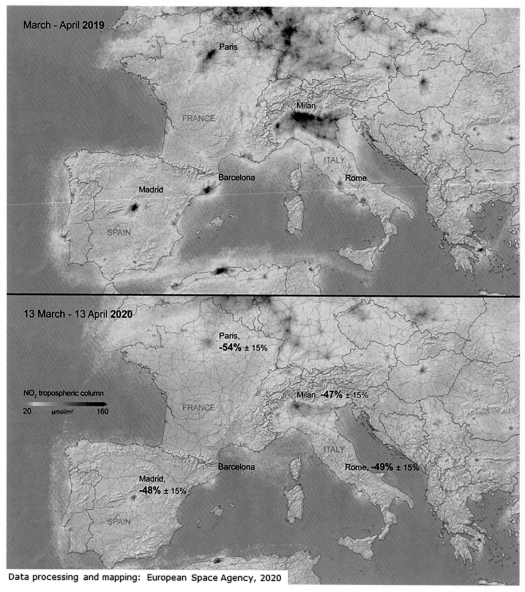

FIG. 4.2   Europe: different dioxide concentrations before and during restrictions for Covid-19 pandemic. *Source: European Space Agency, Nitrogen Copernicus Sentinel-5P satellite.*

for several years.[c] This has accordingly led to a progressive adjustment of the maximum legal limits for pollutant concentration.[d]

As we shift from a European down to a national level, we find it interesting to address the maps produced by ISPRA for regional capitals throughout Italy (ISPRA, 2018). Once again with reference to nitrogen dioxide, analysis (Fig. 4.3A) highlights province capitals that (in the 2017 calendar year) exceeded the annual legal limit for NO_2. Poor air quality areas are clearly found to correspond either to metropolitan clusters (Milan, Rome, Naples, Palermo, Bari, etc.) or to territories whose morphological features strongly affect atmospheric stagnation or air change (most notably Bolzano, Lecco, La Spezia).

With regard to Lombardy, the main provincial capitals (Milan, Brescia, Bergamo) exceed legal limits. Exceedance is also recorded in smaller towns (Lecco, Pavia).

Continuous mapping of the Lombard territory (as per Fig. 4.4) shows that the aforementioned levels of pollution recorded in the capitals expand to dense metropolitan areas and along the major, densely urbanized infrastructure corridors.[e]

Mapping of the Lombardy region related to the 2019 annual average of nitrogen dioxide shows the highest values ($> 40 \, \mu g/mc$) for the metropolitan area of Milan along with the connected conurbations of Brianza, Comasco and Varesotto. Maximum values are also found along the corridor that drives along the A4 Milan-Bergamo-Brescia motorway. The situation is territorially more contained, but still quite significant, for the urban clusters of Pavia, Cremona, and a median stretch of the Camonica Valley.

With pollutant values just below those of dense metropolitan areas, peri-urban clusters outline a geographical map with pollution values from 30 to 40 $\mu g/mc$. Even though settlement density in periurban areas is undoubtedly lower than in dense metropolitan clusters,

[c] "Statistical analysis of 2008–2017 trends carried out on a representative sample of monitoring stations highlighted a statistically significant trend towards reduction of PM_{10}, $PM_{2.5}$ and NO_2 concentrations in urban areas. The slow reduction in the levels of these pollutants in Italy, consistent with what has been observed in Europe over the last decade, is the result of the joint reduction of primary particulate emissions, nitrogen oxides and the main precursors of secondary particulate matter (sulfur oxides, ammonia in addition to the nitrogen oxides themselves)" (ISPRA, 2018, p. 375).

[d] Italian legislation defines legal limits for pollutants in relation to human health in Legislative Decree 155 of 2010:

- for PM_{10} the legal limit as a daily average is 50 g/m; daily limit exceedance is set to a maximum of 35 times. The average annual legal limit is 40 g/m;
- for $PM_{2.5}$ the annual average limit value was 25 g/m until 2019. From 2020 (phase II), it was reduced to 20 g/m;
- for NO_2 the maximum hourly average value is 200 g/m, with an exceedance limit of 18 times a year; the average annual legal limit is 40 g/m.

[e] Mappings relating to pollutant spread are made by drawing on the municipal estimates database made available by ARPA Lombardia (Agenzia Regionale di Protezione dell'Ambiente—Regional Environmental Protection Agency). Values are obtained through municipal aggregate values calculated from the results of simulations on a regional scale carried out with a chemico-physical air-quality model. Data may be accessed on the ARPA portal at https://dati.lombardia.it/. The model was adjusted to 2020 municipal administrative boundaries (for a total number of 1506 municipalities).

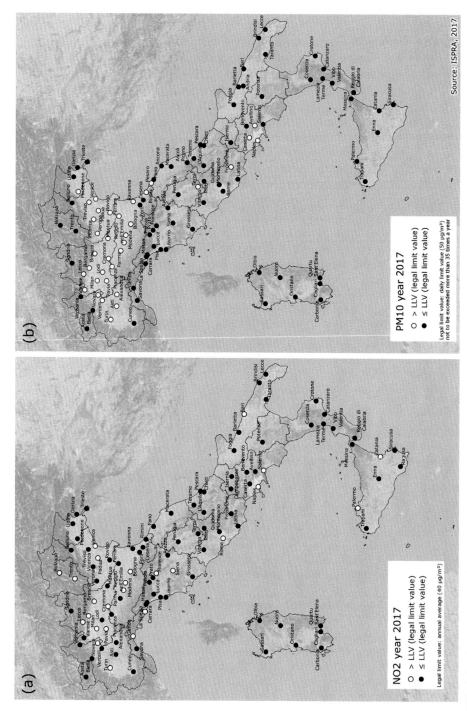

FIG. 4.3 (A) NO₂: annual legal limit exceedances in urban areas; (B) PM₁₀: monitoring stations and daily limit exceedances for health protection.

NO2 year 2017

○ > LLV (legal limit value)

● ≤ LLV (legal limit value)

Legal limit value: annual average (40 μg/m³)

PM10 year 2017

○ > LLV (legal limit value)

● ≤ LLV (legal limit value)

Legal limit value: daily limit value (50 μg/m³)
not to be exceeded more than 35 times a year

Source: ISPRA, 2017

Annual average μg/m³ and number of interested municipalities

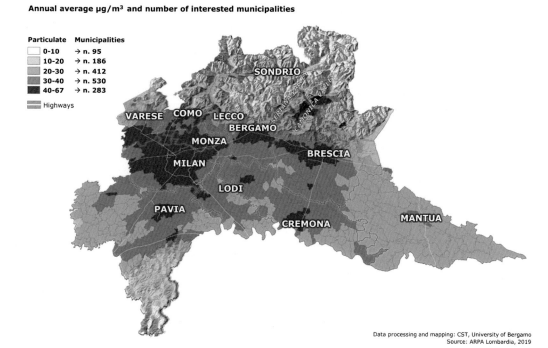

Particulate	Municipalities
0-10	→ n. 95
10-20	→ n. 186
20-30	→ n. 412
30-40	→ n. 530
40-67	→ n. 283
Highways	

Data processing and mapping: CST, University of Bergamo
Source: ARPA Lombardia, 2019

FIG. 4.4 Lombardy: distribution of nitrogen dioxide in the atmosphere in 2019.

periurban clusters present marked territorial relationality. In terms of interdependence, they de facto compete with metropolitan networks.[f]

Color coding in the map extends to entire municipal areas and highlights the predicament of particularly large municipal territories, most notably mountain communities (evident here for the Camonica Valley).

Atmospheric particulate matter (PM_{10} and $PM_{2.5}$) is most often addressed in current studies and research, which aim to trace potential links between the effects of atmospheric pollution and the spread and viral load of SARS-CoV-2 infection.

ISPRA reports (ISPRA, 2018) once again clearly show the distribution of urban clusters that recorded PM_{10} limit exceedances in 2017 (Fig. 4.3B). The specific atmospheric configuration of the Po Valley region is conspicuous for the constant exceedance of legal limit values in all its towns. Nonetheless, if trends recorded in recent years are taken into account, a gradual decrease in pollution values generally recorded may be seen, with occasional variations over the years that in fact turn out to be quite significant.

We now turn to address the scenario of atmospheric particulate matter in the Lombardy region.

[f]See Chapter 3 of this volume on territorial mobility and commute forms.

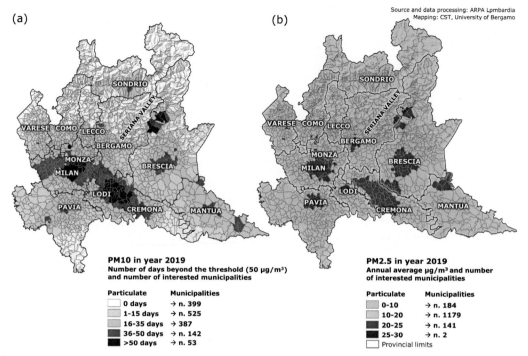

FIG. 4.5 Lombardy: atmospheric pollutants and allowed daily limit exceedances for municipalities.

The topographic map (Fig. 4.5A) is color-coded in red and purple for municipalities that exceeded the threshold of 50 μg/m³ of PM_{10} in 2019 (EU-directive threshold to be contained within 35 days/year).[g] More specifically, municipalities color-coded in purple exceeded the legal limit of 50 μg/mc for over 50 days during the year 2019.

There emerges an extensive territory affected by high levels of exceedances: an area that affects Milan and a substantial part of the Milanese metropolitan area, and which extends south-east to a wide corridor of municipalities in the Lodi and Cremona areas. Along with a few smaller, specific cases (e.g., the towns of Pavia and Sondrio), the urban area of Brescia, a middle section of the Camonica Valley, and two areas in the Mantua region are conspicuous for their extension. It should be kept in mind that color-coding on the map, extending as it does over each entire municipal surface, tends to emphasize municipalities with extensive surfaces, as evident here for Camonica Valley. In such high-density areas, it seems reasonable to connect the pollution levels calculated by the territorial model back to the communities of the valley floor.

Topographic map on the 2019 annual average of $PM_{2.5}$ particulate matter (Fig. 4.5B).[h] In a limited number of cases values are higher than the legal limit of 25 μg/mc, in accordance

[g] As for the map on the number of days per year over the PM_{10} legal limit of 50 μg/mc, we considered the highest value between the PM_{10} and Urbanized PM_{10} for those municipalities belonging to area D of the new regional zoning (105 municipalities along the valley floor).

[h] As for the 2019 annual average map on nitrogen dioxide, a mean value was calculated between NO_2 and Urbanized NO_2 values for those municipalities belonging to area D of the new regional zoning (105 municipalities along the valley floor).

with the legal limit enforced until 2019. Starting in 2020, the legal limit was in fact set at 20 μg/mc. A territorial configuration similar to the spread of PM_{10} may be observed, with the emergence of the Milan-Lodi-Cremona route and the urban areas of Brescia and Pavia and two areas in the upper Mantua region and Camonica Valley. These areas color-coded in red fall within the legal limits established up to 2019. However, since they do exceed the legal limit for 2020, they should be monitored with care to make sure legal requirements are met.

With respect to PM_{10} diffusion, greater distribution uniformity may be observed, with most of the regional territory showing a PM_{10} level between the values of 10 and 20 μg/mc. Readings fall below 10 μg/mc as we get to the pre-alpine and alpine mountain ranges and the Oltrepò Pavese hills.

The map (Fig. 4.6) is constructed on the basis of an index that represents the sum of the "saturation" level of the two particulate components PM_{10} and $PM_{2.5}$ relative to established legal limits (annual average, equal to 40 μg/mc for PM_{10} and 25 μg/mc for $PM_{2.5}$). Achievement of the legal limit reading is therefore equivalent to an index with value 1. Representation via anamorphosis expands regional territories in relation to resident populations. This representation mode makes it possible to immediately grasp the size of the population exposed to pollution sources: in this case atmospheric particulates.

Unlike previous maps, a map which traces cumulated particulate readings effectively shows correspondence between higher-index areas (color-coded in purple and red) and territorial setups with higher urbanization rates.

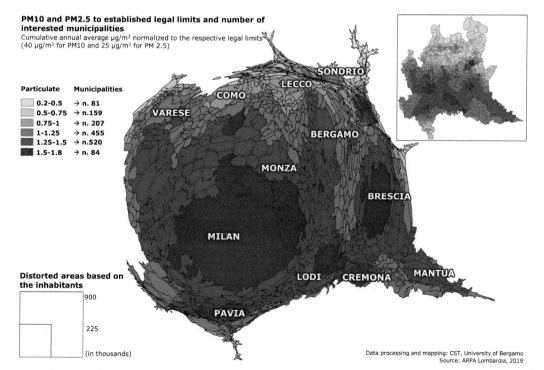

FIG. 4.6 Normalized cumulative indicator of 2019 annual average readings for PM_{10} and $PM_{2.5}$, relative to established legal limits.

While the critical placement of capitals such as Milan, Brescia, and to a lesser extent Pavia, Cremona, and Lodi, may be seen once again in specific maps referring to PM_{10} and $PM_{2.5}$, an integrated version of the same representation introduces additional features of unique interest: significant criticality seems attached to the metropolitan area of Milan, reaching as far as the densely populated areas along foothills, in particular along the Varese and Bergamo routes.

As we move to consider the production system, namely the Companies at Major Accident Risk-RIR[i] and subject to Integrated Environmental Authorization-AIA[j] which constitute environmental pressure factors, we cannot fail to outline a major production and manufacturing system centered around the province of Bergamo.

Analysis of the provincial territory of Bergamo (Fig. 4.7) is carried out by mapping major businesses in terms of their environmental impact. Such businesses are subject to specific measures for containing atmospheric emissions (Companies subject to Integrated Environmental Authorization—AIA), and to major accident risk control protocols (Companies at Risk of Major Accident-RIR).

This systematic mapping is placed on a basemap which reproduces the territory's morphology and provides a functional cognitive background for tracing correlations with the pollutant distribution mapping conveyed in the ARPA model (ARPA Lombardia and Regione Lombardia DG Ambiente, 2020).

Below is an example of the distribution of AIA and RIR companies on a map showing average annual NO_2 readings for the territory of the province of Bergamo as a basemap.

RIR companies on the map are identified with a star symbol, while AIA companies are marked with a circle. Municipality perimeters in the Seriana Valley are mark areas featuring major production facilities and businesses, as widely acknowledged, and documented also in this volume. Also, these are areas severely affected in the onset phase of the Covid-19 epidemic.

The high density of AIA and RIR clusters in the Bergamo area and their significant link with PM_{10} levels in the atmosphere should be noted. It should also be pointed out that these crucial production enterprises in the valleys and highland areas are often located in the densest and most important production clusters, which are located in the wider and more

---

[i]Major accident risk is linked to factories which, due to the presence of dangerous substances in given quantities, may have (an albeit low) chance of causing a major accident in terms of damage to people, property, or the environment. These factories are divided into two groups, those with a lower threshold and those with an upper threshold, on the basis of the amount of dangerous substances they hold relative to established limits set by relevant national legislation. These are usually large industrial plants (such as chemical or oil industries) but phytosanitary warehouses, distilleries and galvanization plants would also fall in this category, given the environmental damage that the substances held in these factories may cause in the event of an accident.

[j]AIA is the authorization measure some companies need in order to comply with the principles of integrated pollution prevention and control (IPPC) dictated by the European Union with directive 2010/75/EU.
A company falls within the AIA procedure when its features meet the requirements of national competence (Annex XII) by Legislative Decree 152/06 and subsequent amendments, and regional (Annex VIII) (plants of regional competence).
The following industry categories are identified by AIA: energy business, metal production and processing, industry of mining products; waste management, paper mills, farms, slaughterhouses, food industries, tanneries, and other activities.

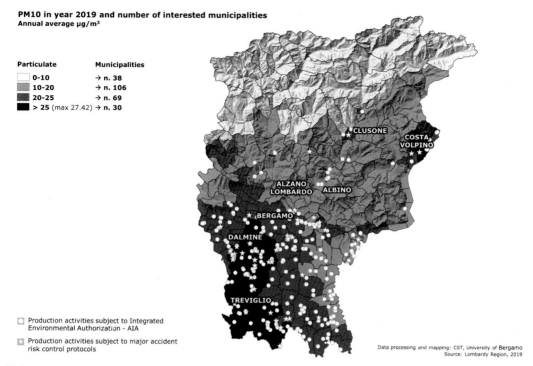

PM10 in year 2019 and number of interested municipalities
Annual average µg/m³

Particulate	Municipalities
0-10	→ n. 38
10-20	→ n. 106
20-25	→ n. 69
> 25 (max 27.42)	→ n. 30

☐ Production activities subject to Integrated Environmental Authorization - AIA

✠ Production activities subject to major accident risk control protocols

Data processing and mapping: CST, University of Bergamo
Source: Lombardy Region, 2019

FIG. 4.7 Distribution of RIR and AIA businesses and annual average readings of PM$_{10}$ in the province of Bergamo.

accessible sectors of the valley bottoms. As for the south-western portion of the flat land, which has the highest NO₂ readings (>µg/mc), a significant density of AIA-RIR companies may be observed, attributable in large measure to the presence of large companies in the chemical sector.

4.1.3 Initial assessment

The correlation between air pollution and contagion, by now acknowledged by a vast literature, was investigated on the basis of two hypotheses that have emerged in the disciplinary debate: the first, worthy of further study, assumes pollution as a Covid-19 diffusion factor by attributing to fine particles (PM$_{10}$ and PM$_{2,5}$) both the role of air-born substrates and the role of viral vectors by way of a "carrier effect." The second hypothesis, long examined in the literature, is based instead on atmospheric pollution as a key component of territorial health, i.e., as a background environmental condition that affects the health of exposed populations.

It should be stressed that original territory analyses conducted in detail on the region of Lombardy must be seen as assessed against existing shortcomings in the air quality detection and monitoring network, against the shortage of detection points and against the unavoidable degree of approximation which territorial data modeling entails (despite our careful application of state-of-the art methods and territorial modeling software, in this case developed by the Lombardy Regional Environmental Protection Agency).

Information to be obtained from the maps with regard to the presence of atmospheric pollutants under investigation (namely PM_{10}, $PM_{2.5}$, NO_2) is therefore reasonably reliable at the regional scale and for large sub-regional territorial fields, but accuracy decreases once we zero in on local scale details.

A reading of the territorial outlines determined by the diffusion of pollutants, some basic considerations regarding distribution may be advanced. First, we should underline a densely urbanized strip of territory in the highlands at the foot of the Lombard pre-alpine and alpine reliefs, characterized by higher levels of fine particulate and nitrogen dioxide pollution, both relative to the context of the Lombardy region, and to the wider national level. Second, a critical line may be observed along a NW-SE Milan-Lodi-Cremona direction, as regards the dispersal of fine particulates. This may be attributable also to the range of production clusters linked to industry, agriculture, and farming. Finally, the distribution of businesses with sizable impacts in terms of atmospheric emissions (in relation to territorial morphologies and atmospheric conditions) may provide a useful indicator for examining territorial exposure to pollutant dispersal.

It goes without saying that any substantive reflection on Lombardy must be placed in the wider context of Po Valley region which may be reaffirmed to present two notable conditions: (i) a high overall density of settlement clusters (in particular due to the richness of the production system) and consequently a high level of atmospheric emissions; (ii) closed-valley morphological and environmental features which facilitate pollutants stagnation in the basin's air mass.

References

ARPA Lombardia, Regione Lombardia DG Ambiente, 2020. Analisi preliminare della qualità dell'aria in Lombardia durante l'emergenza Covid-19. marzo,.

Coker, E.S., et al., 2020. The effects of air pollution on Covid-19 related mortality in Northern Italy. Environ. Res. Econ. 76, 611–634.

Cori, L., Bianchi, F., 2020. Covid-19 and air pollution: communicating the results of geographic correlation studies. Epidemiol. Prev. 44 (2–3), 120–123.

European Environment Agency-EEA, 2019. Air Quality in Europe, 2019 Report.

ISPRA, 2018. XIV Rapporto Qualità dell'ambiente urbano.

Setti, L., Passarini, F., De Gennaro, G., et al., 2020a. Potential role of particulate matter in the spreading of COVID-19 in Northern Italy: first observational study based on initial epidemic diffusion. BMJ Open 10, 1–9, e039338. https://doi.org/10.1136/bmjopen-2020-039338. .

Setti, L., et al., 2020b. SARS-Cov-2RNA found on particulate matter of Bergamo in Northern Italy: first evidence. Environ. Res. 188, 1–5. https://doi.org/10.1016/j.envres.2020.109754.

Yao, Y., et al., 2020. Association of particulate matter pollution and case fatality rate of Covid-19 in 49 Chinese cities. Sci. Total Environ. 741, 1–5.

Chapter 5. Dynamics of contagion and fragility of the healthcare and welfare system

CHAPTER

5.1

Epidemic onset and population and production density of outbreaks in Lombardy

Andrea Brambilla

5.1.1 Premise

The most relevant epidemiological feature of the Covid-19 infection is its frightening propagation speed.[a] It is precisely the virus' ability to spread rapidly across a considerable number of territories which has led us to inquire whether its outbreaks and propagation may be promoted by territorial features such as population density or functional connections.[b]

In fact, first-hand data culled from the Italian media seem to suggest that diffusion outbreaks may be triggered by mobility and contacts between inhabitants of the same region, whose commutes bring them into contact with other inhabitants from the same region or neighboring ones. At a later stage, it would seem that outbreak diffusion occurs by proximity and therefore through proximity contacts, unless the outbreak is stemmed.[c]

---

[a] As epidemiologist Pierluigi Lopalco, professor of public health at the University of Pisa, pointed out, "what should concern us in a pandemic is less overall lethality than the speed whereby cases are spread" (March 1, 2020: https://www.adnkronos.com/fatti/cronaca/2020/03/01/coronavirus-paura-letalita-velocita-diffusione_75XoZwJ9lnFf30ko9RfLbI.html).

[b] This section adopts the theoretical-methodological approach illustrated in the introduction to the present volume, namely in Casti's discussion of the phase defined as the "onset phase" of the Covid-19 epidemic. The charting of outbreaks and the tracing of possible ties between them and the production and logistics network is based on this initial discussion. For an in-depth treatment of this issue, see also Casti (2020).

[c] This is the case of the Covid-19 outbreak in the municipalities of Nembro and Alzano Lombardo. On March 3, 2020, the Italian Technical and Scientific Advisory Group had strongly advised to extend the same restrictive measures already adopted for red zone municipalities around the town of Codogno to these two municipalities, in order to limit the spread of infection to neighboring areas. The group's minutes were published by the local newspaper L'Eco di Bergamo on August 7, 2020 at the link: https://www.ecodibergamo.it/stories/premium/Cronaca/mancata-zona-rossa-ecco-il-verbalealto-rischio-a-nembro-and-alzano_1367162_11/.

In the Po Valley,[d] Covid-19 outbreaks are located in the vicinity of metropolitan or urban areas (examples are Codogno, Alzano Lombardo and Nembro in Lombardy, Sale and Tortona in Piedmont, Vo' Euganeo in Veneto and Medicina in Emilia-Romagna). Such areas, however, belong to a much wider expanse of polycentric urbanization that characterizes this macro-region.

The study presented here provides first a broad outline of this layout, by tracing sporting event links between the first two Lombardy outbreaks and then looking at the production dynamics of their respective territories, which is seen to operate both nationally and internationally (Casti and Adobati, 2020a,b). Subsequently, analysis is extended to other outbreaks of the Po Valley with the aim to discuss their diverse features.[e]

5.1.2 Outbreaks and sporting events in Lombardy

Possible links between the two outbreaks of Nembro-Alzano Lombardo (in the province of Bergamo) and that of Codogno (in the province of Lodi) were initially investigated on the basis of sporting events.[f] Such events indicate that contacts between the two outbreak areas were intense, which lends substance to the hypothesis that contagion must have occurred through reticularity rather than "proximity."

Before modeling this sport-related reticularity between Val Seriana and the Codogno area, we need to briefly recall the sporting event itself, which several sources pinpointed as the initial trigger for the epidemic onset that was to affect the area of Bergamo most severely. It was the Atalanta-Valencia match played on February 19, 2020 at the San Siro Stadium in Milan. This Champions League match, which was to hit major headlines nationally and internationally, highlighted the fact that, in times of pandemic, public events inevitably lead to crowding, which determines the onset of an outbreak, especially when, as in this case, the turnout is very high. More than 40,000 football supporters from Bergamo and Valencia are reported to have

[d]The Po Valley is the largest Italian plain. With nearly 20 million inhabitants, this is the most densely populated area in Italy as well as among the densest (and most polluted) areas in Europe. In 2000, Italian geographer Eugenio Turri famously named this densely inhabited and urbanized macro-region as the Padan Megalopolis (*Megalopoli Padana*); see: Turri (2000).

[e]With the exception of the two Chinese tourists who tested positive on January 30, 2020, the first full-blown case of Covid-19 in Italy was recorded on February 21 in the town of Codogno, a Lombard municipality located in the province of Lodi, about 65 km south of Milan. Over the next few days, the province of Bergamo, also in Lombardy, recorded the most dramatic surge in Covid-19 cases, particularly in two municipalities of the Seriana Valley (Alzano Lombardo and Nembro), located 15 km northeast of the province city of Bergamo. Quantitative data on the Covid-19 epidemic in Italy may be found on the Italian Department of Civil Protection (*Protezione Civile*) website at: http://www.protezionecivile.gov.it/.

[f]On February 23, 2020, following a Decree issued by the Italian government, the Lodi-area outbreak was to become the first "red zone" in Italy and Europe. For a space–time analysis of the political measures adopted by the Italian government to deal with the Covid-19 contagion, please refer to Chapter 6 of this volume. You may also wish to consult the legislative decree itself, or DPCM—Decree of the President of the Council of Ministers—dated February 23, 2020 at the following link in the Italian Official Gazette: https://www.gazzettaufficiale.it/eli/id/2020/02/23/20A01228/sg.

taken part in the event.[g] That an event of this magnitude could speed up Covid-19 infection between supporters of either team comes as no surprise. Local news recorded the festive atmosphere that preceded the match, given that it marked Atalanta's first appearance in the round of 16 for the most prestigious European football competition, the Champions League. Teamwork, rather than outright rivalry, was said to mark the event.

Starting from this incident, we set out to investigate sporting events, not so much with the aim to chart their occurrence in the Lombardy region but rather in order to: (1) investigate whether there existed a reticularity model between such events which would indicate close links; and (2) assess whether such reticularity might also involve the Italian provinces which would later record a significant number of cases, namely the province towns of Lodi and Bergamo.

We sifted through sports events of varying sizes across different categories over the time period from February 1 to 22, 2020, on which date an ordinance of the Lombardy Region de facto prohibited all competitive and non-competitive sports events.[h] Our aim was to investigate outbreaks territorially and to establish whether sports events did impact local outbreaks, since such events openly promoted large-scale gatherings between members of different communities.

The first event we addressed was a football match: precisely the B-round Lombardy Excellence match,[i] which featured teams from the provinces of Bergamo, Lodi, Lecco, from the eastern area of the provinces of Milan and Monza and Brianza as well as from the northern area of Cremona. The Real Casal Codogno 1908 Football Team, based in the municipality of Codogno, in the province of Lodi, the very first Covid-19 outbreak center in Italy, also took part in the competition. A long series of sports events involving these territories may be cited to uphold our initial hypothesis of reticularity and interrelation, namely:

The match took place on February 9, 2020:

- At the Falco Stadium in Albino—located in the province of Bergamo, 14 km north-east of Bergamo—AlbinoGandino played RC Codogno 1908, with an estimated presence of about 250 people residing in the territories of Val Seriana or the Codogno area.[j]

---

[g] Atalanta played in the Giuseppe Meazza San Siro Stadium (Milan) because the Gewiss Stadium in Bergamo was being refurbished and could not host UEFA Champions League matches for safety reasons.

[h] See Ordinance dated 22 February 2020, signed by the President of the Lombardy Region Attilio Fontana, to be found at: https://www.regione.lombardia.it/wps/portal/istituzionale/HP/lombardia-notizie/DettaglioNews/2020/02-febbraio/24-29/specifica-ordinanza-firmata-da-fontana.

[i] The "Lombardy Excellence" match, which ranks fifth in the national Champions League. The event ranks second by importance at the amateur level and is the most significant event on a regional scale. Lombardy sets up three rounds of 16 teams each. Specific match data may be consulted on the website of the FIGC-Italian Football Federation National Amateur League of the Lombardy region at the following link: www.crlombardia.it/.

[j] Attendance figures for all football and volleyball events listed in this study are estimated. Since no official attendance numbers were released, figures were estimated by cross-referencing data such as the number of athletes involved, seasonal attendance averages and participation statistics for analogously classed events over previous years. Data were culled from online and offline media sources.

On February 16 two matches were played:

- At the Molinari stadium in Codogno—in the province of Lodi—RC Codogno 1908 played Mapello, a team from the homonymous town located approximately 15 km northwest of Bergamo, with an estimated presence of around 200 people.
- At the Municipal Stadium of Verdellino—located in the province of Bergamo, about 21 km from Bergamo itself—ZingoniaVerdellino played AlbinoGandino in a match which mingled supporters from the municipalities of Verdello and Zingonia, in the lower plains around Bergamo, with those from the Lower Seriana Valley, with an estimated presence of about 200 people.

Another team from the Lodi province also took part in this championship, the Sant'Angelo team from the Lodi municipality Sant'Angelo, located close to the Codogno "red zone," at a distance of around 30 km. More specifically, on February 9:

- Sant'Angelo played Cisano at the CS Cisanese Stadium in Cisano Bergamasco—a Bergamo municipality located 19 km northwest of the Orobic capital—an event which, once again, brought supporters from Bergamo and Lodi into close contact.

Bocce ball tournaments were another class of events our study addressed.[k] In this instance, On February 9:

- The final round of a key Bocce Ball Tournament named "Raffa" specialty, valid at a national level, took place at the Ranica Bocce Ball facility, located in the province of Bergamo, 6 km from Bergamo, in the Lower Seriana Valley. This tournament was also attended by a team from Codogno. Tournament preliminaries took place in various other facilities, in addition to the one in Ranica. Such bocce ball facilities were located in the province of Bergamo and precisely in the municipalities of: Bergamo, Zogno, Urgnano, Montello, Dalmine, Presezzo and Credaro. Athletes for the Codogno team played an excellent tournament, ranking among the first and playing numerous matches with multiple Bergamo teams.

The third sporting category our study investigated in order to trace connections between reticularity and sports events involving the Codogno area and the Seriana Valley was athletics.[l] In this case,
On February 2:

- The Bergamo city Marathon took place. It ran inside the city of Bergamo and involved more than 1250 athletes, mostly from Lombardy, as well as another 489 athletes for minor competitions, such as the 10 km run or the relay race.

The final category we investigated was volleyball,[m] with regard to two key matches which shed light on our model of geographic reticularity:

---

[k]Data were culled from the FIB-Italian Bocce Federation website at: https://www.federbocce.it/.

[l]Data were assembled from the Bergamo City Run website, available at: https://www.endu.net/it/events/bergamo-city-run-dei-mille/.

[m]Data sources in this case were the sites of the Italian Volleyball Federation and of the Regional Committee of Lombardy, which may be consulted respectively at the links: www.federvolley.it/ and www.lombardia.federvolley.it/.

On February 5:

- At Pala Gossolengo in Gossolengo, in the province of Piacenza in the south-west area of Codogno, an Italian First-Division Womens match was played between the teams BusaFoodlab Gossolengo and Warmor Volley Gorle, the latter based in the homonymous Bergamo municipality, about 6 km from the city of Bergamo and overlooking the Val Seriana, with an estimated presence of about 250 people.

On February 8:

- A Second-Division Mens match took place at the Pala Bertoni in Crema between the Imecon of Crema and the Volleymania of Nembro, respectively located near Codogno and at the very heart of the Seriana Valley (the first outbreak site for the area of Bergamo) together with Alzano Lombardo, with an estimated presence of around 300 people.

Both matches involved teams coming respectively from the Seriana Valley (Nembro and Gorle) and from areas adjacent to the first "red zone," namely Gossolengo in the province of Piacenza and Crema in the province of Cremona. It should be noted that the latter localities were among the first provinces to report a substantial number of infections, after the province of Lodi.

It may therefore be concluded that in the period immediately preceding the "onset" phase of Covid-19 contagion, a series of sporting events involved and intertwined Codogno and the neighboring municipalities with the Bergamo and Val Seriana territories, as highlighted in Fig. 5.1. This analysis clearly shows a very dense reticularity across the Po Valley reticularity. It also suggests that provided one thinks of an outbreak site less in terms of a zero-index model (the "zero-patient" first infected by Covid-19) than in terms of a Covid-19 pathogenic area of contagion which may threaten an entire community (namely the Codogno outbreak), one reaches the conclusion that, even though it was promptly addressed by the medical authorities, the Codogno outbreak was far from isolated. If anything, at the very onset of the Covid-19 disease, Codogno was reticularly connected with areas of the lower Seriana Valley, which were to record the second main Covid-19 outbreak in Italy.

The relevance of a reticularity model is further corroborated by charting the epidemic's temporal trend: contagion data dating back to 2 March,[n] 12 days after the first confirmed Covid-19 case in Codogno, attest that in the province of Bergamo 209 people had already tested positive for Covid-19. Of these, 87 were in the Seriana Valley.

The use of reflexive cartography (Fig. 5.1) to visualize the space–time evolution of sport-event reticularity between the Codogno and Seriana areas in February 2020 underlines, on the one hand, the multiplicity of sites where sports events were held. On the other hand, reflexive cartography (Casti, 2015) also highlights the multiple links between these territories, determined by the flows of participants which are marked on the map by means of arched arrows in black. While such data readings necessarily come short of fully accounting for the convoluted dynamics of Covid-19 contagion, they effectively chart the complex interplay of contacts

[n] Data source: https://www.bergamonews.it/2020/03/02/coronavirus-il-punto-a-bergamo-e-in-provincia-e-le-limitazioni/357146/.

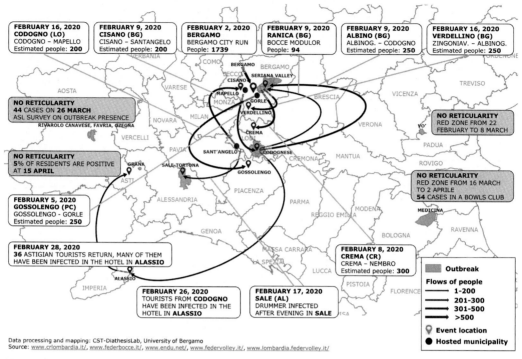

FIG. 5.1   Reticularity between outbreaks in Northern Italy.

and ultimately underline the impact of mobility, with regard to those involved in such sporting events, either as athletes or spectators. The resulting outline records a densely networked and dynamic territory.

5.1.3 The production and logistics fabric underlying Lombardy's outbreaks

Before addressing the other Po Valley outbreaks, we should consider the production and logistics fabric that lies under Lombardy's outbreaks. The goal is to pinpoint features that integrate our analysis of the dynamic and the reticular layout of this regional area. Both the territories around Codogno in the province of Lodi and the area of the Seriana Valley in the province of Bergamo feature a highly developed productive fabric that relies on a dynamic logistics system. Production and logistics are favored by proximity to two major road networks for Italy and Europe: the A1 Milan-Naples motorway (route E33 in the European classification) and the A4 Turin-Trieste motorway (routes E64 and E70 in the European classification).

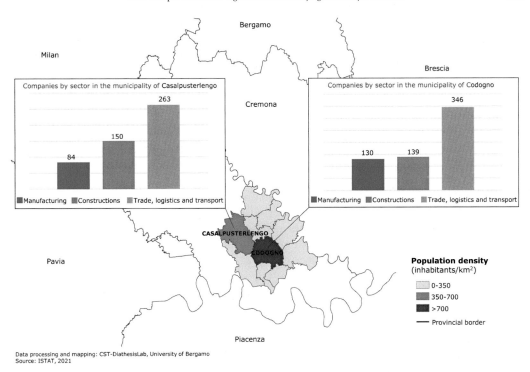

FIG. 5.2   Population and companies density in the province of Lodi (Codogno and Casalpusterlengo).

As regards the province of Lodi, our map (Fig. 5.2)° shows that the territory of the municipalities of Codogno and Casalpusterlengo, the two main centers belonging to the first "red zone," was marked off as a high-infection zone by legislative decree on 23 February 2020. What we have is on average a high-production area, as evidenced by the number of companies attached to the manufacturing, construction, and trade sectors, both retail and wholesale,

°The map was created using the Lombardy Region Geoportal database. The municipalities included in the first "red zone" were color-coded according to the number of inhabitants. The graphs were plotted using the ISTAT database (available at: http://dati.istat.it/Index.aspx?DataSetCode=DICA_ASIAUE1P), which classifies employees by economy sector. Companies classified as manufacturing businesses include all types of manufacturing, from the textile to the food industries, as well as businesses belonging to sectors such as mining, electricity or gas supply, and waste management. The construction industry comprises all the businesses related to the building and civil engineering sectors, while the trade, transport and logistics industries are taken to include all the businesses that involve trade, both retail and wholesale, transport, logistics, and warehousing. The datasets we addressed refer to the 2018 calendar year and are based on the ATECO code: a classification system adopted by ISTAT for national statistical surveys related to economics.

and to logistics. There, most companies have between 1 and 9 employees.[P] However, medium and large enterprises are also present. Even a cursory assessment of the local economy fabric shows that manufacturing is crucial both for the local economy and for its reticular connections. And of course, the latter may have contributed to the swift spread and intensity of the Covid-19 outbreak in the area.

The productive fabric around Codogno is remarkably nimble: it hosts two of the most prominent businesses in the Lodi area, namely Pellinindustrie, a technical shading system manufacturer, and MTA, a manufacturer of electrical and electronic components for the automotive market, with a major branch in the city of Shanghai, China[q] which indicates the local presence of an industry-driven reticularity tied to the import–export business.

Another productive sector of the Codogno area is agriculture. The Autumn Livestock Fair takes place every year in November.[r] That is also why Codogno ranks among the top municipalities of the Lodi agricultural district. Codogno also hosts the "Tosi" Agricultural Technical Institute, attended by students from other provinces and regions, such as the student from Valtellina who first tested positive for Covid-19 in the province of Sondrio.

Data from the Chamber of Commerce of Milan, Monza and Brianza and Lodi recording exports to the Lodi province in 2019 confirm this first "red zone" area plays a prominent role. A report on import and export activities for this region is worth citing in full to gain an overall idea of the area's economic relevance (Camera di Commercio di Milano and Monza e Lodi, 2020, pp. 13–14).

In 2019, Lodi companies exported goods for an aggregate value of nearly 3.6 billion euros. Electronics is clearly the leading sector, worth nearly 1.4 billion euros in exports, that is 38.9% of the province's overall amount. The second sector is chemistry, with a total amount of 586 million euros, while the third sector is food. Three other sectors exceed the threshold of 100 million euros for exports in 2019: electrical appliances, machinery and rubber-plastic. Imports in 2019 reached an aggregate value of 6.6 billion euros, or 3 billion more than exports. Import of electronic goods is worth 2.3 billion euros, which corresponds to 34.8% of the overall imports for the Lodi area. In fact, most of the imports for Lodi business, that is over 70%, refers to only three sectors: in addition to electronics, the other two are pharmaceuticals and food products, for an amount close to 930 million euros. Exports of goods from the Lodi area decreased by 2.6% in 2019.

If we consider data related to the fourth quarter of 2019, nearly all of the over one billion euros generated by Lodi exports are tied to European markets: that is over 965 million euros, equal to 92.7% of the value of exported goods. Outside Europe, nearly 43 million euros in goods are shipped to Asia, 18 million to America and nearly 14 million to Africa.

[P] Throughout history, small and medium enterprises (SMEs) in Italy have always played a leading role: "Small and medium-sized enterprises, defined here as active enterprises with a turnover of <50 million euros, employ 82% of workers in Italy (well above the EU average) and represent 92% of enterprises. These figures show that SMEs are a key feature of the Italian economy. They reflect traditions and entrepreneurial practices that are well established in these territories." The quote comes from Sole24Ore, the leading business newspaper in Italy, in an analysis of SME data dated 2019. The article may be consulted at: https://www.infodata.ilsole24ore.com/2019/07/10/40229/.

[q] MTA business data may be accessed online at: www.mta.it/home.

[r] Source: http://www.fieradicodogno.com/.

As for imports, about half of the supplies of goods come from Europe: namely 54.3%. The remainder is almost exclusively of Chinese origin: in the fourth quarter of 2019, Lodi companies sourced goods from China for a total of 875 million euros, 44.2% of the province's total imports. The negative exports trend for Lodi fully reflects the performance of European countries (−14.8%), in particular the European Union (−15.2%) which, as mentioned, makes up over 90% of the flow of goods directed abroad. Yet, even exports to other world areas faced a significant drop compared to the fourth quarter of 2018: direct exports to Asia fell by 10.5%, those directed to the Americas by 29.8% (−9.3% for the United States), and those to Africa by 10.2%. Some markets show notable positive changes in terms of percentage growth, for example China (+58.5%) and India (+22.5%). However, these are unremarkable flows in absolute values. As for the imports dynamics, a 10.6% decline is the result of a dynamic close to zero for goods arriving from Europe (+0.6%) and a substantial reduction in imports from Asia instead (−21%), especially Chinese (−21.8%) (Camera di Commercio di Milano and Monza e Lodi 2020, pp. 13–14).

This report shows how exchanges between the Lodi area and China are economically and numerically substantial. Intense contacts between China, where the Covid-19 epidemic originated, and the Lodi area, in which the first Covid-19 outbreak in Italy was recorded, probably occurred.

The province of Bergamo and the Seriana Valley also boast remarkably dynamic productive areas. These are distributed both along the piedmont area, where the A4 motorway runs, and along the valleys. Among the latter, the Seriana Valley stands out as a high-density productive area, where the spread of contagion was in fact most swift and severe. Fig. 5.3, which refers to the Seriana Valley, lists municipalities that were taken into account for an analysis of the productive fabric: namely Albino, Alzano Lombardo and Nembro, the three Seriana towns with a population of over 10,000 inhabitants. Even in this case, small businesses, with a number of employees between 1 and 9, are prevalent. The manufacturing sector features instead the highest percentage of medium-to-large businesses.

An assessment of data collected by the Chamber of Commerce (Camera di Commercio di Bergamo 2020b) shows that the productive fabric of Bergamo is made up of 107,296 businesses, which employ nearly 400,000 people. Of these, 3331 are involved in transport and warehousing,[s] and include a total of 23,780 employees, an increase of 2090 units compared to the same period of the previous year. That marks the most substantial increase, trailed only by that of employees operating in the accommodation and catering services sectors.

Data show how the transport and logistics sector, in addition to being an expanding sector, is remarkably solid in the province of Bergamo. Overall, the map highlights the importance of the lower Seriana Valley in the productive fabric of Bergamo, attested by the wide extension of production areas that chart a seamless line of business activities along the central axis of the valley. Because of their frequent exchanges with outside regions, such businesses may have contributed to the rapid spread of local Covid-19 infection, a fact that underlines the fragility of contemporary living.

---

[s]More specifically, these companies belong to category H of the ATECO classification code for Italian economic activities, adopted by the ISTAT and corresponding to the "transport and storage" activities.

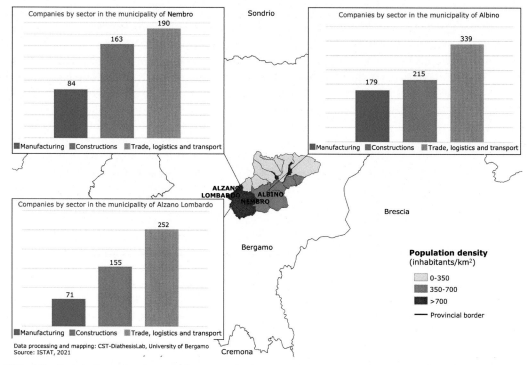

FIG. 5.3 Population and companies density in the Seriana Valley outbreak (Bergamo).

Intense exchanges with outside regions are attested by 2019 exports data drawn up by the Bergamo Chamber of Commerce. These data rank China as the second country (after the United States) outside the European area by exports value.[t] Germany ranks first with 2621 million euros worth of exports (although the value has decreased on an annual basis). And, incidentally, Germany was the first European country to have recorded a Covid-19 outbreak at the end of January.

In conclusion, the two provinces of Lodi and Bergamo, the first to have recorded an alarming increase in Covid-19 cases, are characterized by marked productivity and intense economic exchanges with the economies of China and Germany: with the former in the case of Lodi, and the latter in the case of the Bergamo area. In addition, the province of Bergamo, partly facilitated by a solid road network infrastructure, greatly increased the number of employees in the transport and storage businesses. Such data lead us to assume that, at least in an initial phase, the logistics and transport sector contributed to the swift spread of Covid-19 from China or Germany, by way of the many businesses from those countries which have frequent economic relations with companies in Lombardy.

[t]Exports from the province of Bergamo to China in 2019 amounted to 466 million euros and were up 8.3% on an annual basis compared to 2018. For more information, see Camera di Commercio di Bergamo (2020a).

5.1.4 Reticularity between outbreaks in Northern Italy

To gain a clearer understanding of the reticularity links between regions which had an impact on rapid contagion spread and led to major outbreaks in the Italian territory, we extended our analysis to outbreaks in regions which border Lombardy, namely the areas of Liguria, Piedmont, and Veneto.

The reticular connections thus analyzed, which are visualized in the same map, can be invariably traced back to sporting events that took place at the end of February 2020. Such events chart clear links between the Codogno outbreak in Lombardy and the regions of Liguria and Piedmont. In order to address this reticularity, we used the only sources available: the press and media. In most cases, news media sources attest to a reticularity either involving single subjects or small groups, who caused significant outbreaks in Northern Italy in their movements between regions. The resulting model is a multiple reticularity, that is multiple networks of interconnected outbreaks. Conversely, other outbreaks showed no reticularity and their link with other outbreaks was weak, if not altogether absent.[u]

Connections in this case were different. They charted either networked or isolated outbreak clusters that are going to address in detail for the Veneto, Emilia-Romagna and Piedmont regions, with a focus on the municipalities of Vo' Euganeo in the province of Padua, Medicina in the province of Bologna, Grana in the province of Asti and the municipalities of Rivarolo canavese, Favria and Ozegna in the province of Turin.

As regards networked outbreaks, the regions involved were Piedmont and Liguria. Covid-19 contagion data for June 21, 2020 rank Piedmont second in Italy, after Lombardy, in terms of recorded cases. Proximity to Lombardy was certainly a factor in this context. However, reticular links also affected outbreaks in Piedmont, along with subsequent contagion by proximity. Because of this, in order to analyze the first Piedmont outbreak, recorded in the town Asti, we also need to include the region of Liguria.

For the first Covid-19 outbreak recorded in Liguria was in hotels in the town of Alassio,[v] later tied to the Covid-19 surge in Piedmont. We are dealing in this case with multiple reticularity, because some tourists from the Codogno area, who stayed at this hotel and tested positive, infected in turn a group of tourists from Asti. We would use the term multiple reticularity because the outbreak that developed here—originally from Codogno—in turn spread outside the region, infecting people who lived elsewhere. In particular, we would like

[u] This difference may be found also in our subsequent analysis of outbreaks across the Po Valley. It should also be noted that the economic and social setup of the latter regions, to which these outbreaks belong, differs substantially from the one recorded for network outbreaks: it is based either on the lack of strong links with the main infrastructural network, or on the relative absence of gatherings between members of different municipalities. Data source: ISTAT.

[v] At the Alassio hotel, groups from the Asti region met with groups from the Codogno area, which caused a large Covid-19 outbreak within the hotel; see: M. Calandri, "Coronavirus, ad Alassio nell'hotel del contagio," in: *La Repubblica*, 28.02.2020, website: https://www.repubblica.it/cronaca/2020/02/28/news/coronavirus_ad_alassio_nell_hotel_del_contagio-249843400/.

to point at least two facts: the regions involved are at least three, and they are quite different from each other; a given region may be seen to act as a gathering site for residents of two different regions which are not directly connected otherwise. In this case we are dealing with a hotel. However, the same could have occurred in any sites which generate people flows such as airports or stations.

Even the Covid-19 outbreak in the towns of Sale and Tortona[w] developed inside a facility that serves as a center for residents, albeit only on a supra-municipal scale. It is known that, in spring 2020, Alessandria was one of the most affected Italian provinces in terms of Covid-19 infections relative to its resident population.[x] It was also one of the first provinces to be marked off as an "orange zone."[y]

On March 3, 2020, the mayor of Sale issued a notice[z] aimed at anyone who, starting from February 17, 2020[aa] may have been frequenting the *Comet dance hall* and may now be showing symptoms attributable to Covid-19. They were all invited to contact their GP—General Practitioner. The notice was issued after a drummer who had played at the ballroom on that night tested positive for Covid-19 the next day. A Covid-19 outbreak in the municipalities of Sale and Tortona ensued. It was later found that drummer belonged to a band who had been playing in several clubs across Lombardy, not far from the Codogno area. Also in this case, we would speak of a reticularity model, which involved the Po Valley and its mobile lifestyle, with people and musicians on tour from different areas converging inside a closed space, such as a ballroom, in a gathering that inevitably led to an outbreak onset.

As we turn to consider isolated outbreaks in the Po Valley, that is outbreaks not reticularly linked to other regions, we come across two instances: one in the Venetian municipality of Vo' Euganeo in the province of Padua and the other in the Emilian municipality of Medicina, in the province of Bologna.

---

[w] Tortona is a Piedmont town in the province of Alessandria of about 27,000 inhabitants, located south-west of the Po Valley on the border with the Ligurian Apennines; Sale is a small town of about 4000 which adjoins Tortona.

[x] On June 22, 2020, the province of Alessandria recorded 4058 Covid-19 cases, equal to 0.963% of the local population. At the time, this reading made it one of most severely affected provinces in Italy in terms of the infected people/population ratio, against a national average of 0.395 and a Piedmont region average of 0.717. See: https://lab24.ilsole24ore.com/coronavirus/.

[y] In the night between March 7 and 8, 2020, the Italian Prime Minister issued a decree—DPCM of March 8, 2020—with restrictive measures that applied to Lombardy and 14 provinces of the Center-North (Modena, Parma, Piacenza, Reggio nell'Emilia and Rimini in Emilia-Romagna, Pesaro and Urbino in the Marche, Alessandria, Asti, Novara, Verbano-Cusio-Ossola and Vercelli in Piedmont and Padua, Treviso and Venice in Veneto) for a total of nearly 16 million people. Rather inappropriately, this lockdown was named as the "orange zone," to set it apart from the "red zones" of the Codogno area and Vo' Euganeo. The published decree may be found on the Italian Official Gazette at the link: https://www.gazzettaufficiale.it/eli/id/2020/03/08/20A01522/sg.

[z] The notice is posted on the website of the Municipality of Sale, at: https://www.comune.sale.al.it/it-it/avvisi/2020/protezione-civile/avviso-sindaco-132680-1-748e493c269a1dc82a7d81dda78f31a3.

[aa] The municipal decree mentions the date of February 17, while an Italian Television Report blog article mentions February 14, Valentine's Day. Source: www.report.rai.it.

The first outbreak, in Vo' Euganeo, developed in a municipality of 3305 inhabitants[ab] at the foot of the Euganean Hills, in the province of Padua, within which, on February 21, 2020, the first Covid-19 case for the Veneto region was reported, barely a few hours after the first case was recorded in Codogno.

Unlike other outbreak areas we addressed earlier, such as the Codogno area or the Seriana Valley, Vo' is not directly connected with medium or large road networks.

The municipality of Vo' has become a case study at the Department of Molecular Medicine of the University of Padua, precisely in relation to its geographical and social features, and especially given the fact that no significant reticular links to the other outbreaks developed in Northern Italy have been ascertained (Lavezzo et al., 2020).

The second outbreak was recorded in the municipality of Medicina, in the province of Bologna, in the Emilia-Romagna region. Following 54 confirmed Covid-19 cases in a local bocce ball venue, Medicina, which numbers 16,768 inhabitants, was marked as a "red zone," on March 16, 2020.[ac] The provision lasted until April 3. Unlike Vo', the municipality of Medicina is a medium-sized municipality located near Bologna. However, it has no railway station and the nearest motorway access, that of Castel San Pietro Terme on the A14 Bologna-Taranto motorway, is 10 km away. Unlike the Codogno area or the Seriana Valley, whose economies rely on industry and logistics, the municipality of Medicina is still largely an agriculture-based municipality, albeit in terms of intensive agriculture. Even in this case, no marked reticularity existed with other outbreaks in Northern Italy.

Even within the Piedmont region, two outbreaks were identified in addition to the reticular outbreak of Asti. Neither, however, exhibited strong reticularity with other outbreaks recorded in Northern Italy or more specifically in the Piedmont region. The first outbreak was in the town of Grana[ad] in the province of Asti, while the second developed across three municipalities in the Canavese area: Rivarolo, Favria and Ozegna, in the province of Turin.[ae]

Our survey of Northern Italy (Fig. 5.1) outlines a clear distinction. A number of Covid-19 outbreaks recorded in municipalities that were neither tightly connected with major infrastructure networks nor equipped with tourism, trade or logistics facilities. In these cases, the propagation of the epidemic was generally stemmed, and containment and health

---

[ab] Vo' is a small town which features low reticularity with the surrounding municipalities. Its territory hosts a number of schools, from kindergarten to lower secondary school, but no upper secondary schools, which would involve school commuting through public transport. In this regard, see Chapter 1.4 of this volume, which focuses on contagion in the Bergamo area and addresses factors of territorial fragility that may have impacted outbreak onset in the Seriana Valley.

[ac] By decree of the governor of the Emilia-Romagna Region, Stefano Bonaccini in agreement with the government and the local council. Source: https://bologna.repubblica.it/cronaca/2020/03/16/news/ coronavirus_medicina_e_zona_rossa_vietato_entrare_e_uscire-251412467/, for population data see: istat.it.

[ad] On 15 April, 5% of the 600 inhabitants in Grana swab-tested positive for Covid-19. Many of these resided in the local Nursing and Residential Care Facility (RSA), so the mayor deemed it best to apply for a "red zone" classification. Source: www.lastampa.it/asti/2020/04/15/news/troppi-contagi-alla-casa-di-riposo-grana-deve-diventare-zona-rossa-1.38721120.

[ae] On 26 March, 44 people positive for Covid-19, a reading that was immediately picked up for a statistics survey by the To4 Local Health Authority (ASL). Source: www.lastampa.it/torino/2020/03/26/news/ coronavirus-troppi-contagi-e-tre-comuni-piemontesi-diventano-un-caso-1.38640865.

measures implemented effectively. Conversely, large Covid-19 outbreaks located along the main corridors of territorial mobility made containment measures difficult and sluggish, which in turn engendered high contagion diffusion by proximity. For example, the Covid-19 outbreak at Alzano Lombardo-Nembro, the contagion epicenter in the Seriana Valley in the province of Bergamo, recorded a sharp surge in the number of confirmed infections, from 0 to 87, within the space of a week, as seen in Chapter 1.4.

It can therefore be concluded that, even within dissimilar territorial configurations, the mobile and reticular living patterns of the Po Valley are likely to have determined a diffusion of Covid-19 contagion both by reticularity and by proximity. To be sure, any conclusive statement in this sense requires caution, since the variables involved in the spread of the contagion are hard to monitor: from large-scale and frequent exchanges tied to productivity, trade and logistics within a vast infrastructure network, to local gatherings linked to sporting events or social gatherings. The resulting picture is diverse.

5.1.5 Initial results

The results show strong sports-driven reticularity between the Codogno area and the province of Bergamo and the Seriana Valley in Lombardy, and a first and second level tourism-driven reticularity between Lombardy, Liguria, and Piedmont. Analysis of the logistics and production fabrics of these regions also shows significant international ties with countries affected by Covid-19 (namely China, and Germany, as the first confirmed European outbreak site) and the possible regional spread of Covid-19. On the contrary, the outbreaks that occurred in the neighboring regions differ both in their onset and in their connections, since those are regions located on the margins of the Po Valley megalopolis, and therefore less connected.

This research shows that the Po Valley, in particular in its central part, where population density is high, outlines a region that clearly facilitated the onset of Covid-19 infection. That is arguably due to the region's dynamic production and the intense exchanges between its inhabitants, which lead to gatherings and multiple contacts. These, in turn, may be seen as the distinctive features of the new type of reticular, mobile, and connected living, albeit within a varied gamut of regional differences, as seen above. In this perspective, the cartographic mapping we produced underlines, on the one hand, the space–time evolution of reticularity for these regions, enhanced by various events that took place throughout February 2020. On the other, this map highlights the key-role played by productive activities in the contagion epicenters of the Codogno area and the Seriana Valley, which feature a strong network of economic ties on a global scale.

References

Camera di Commercio di Bergamo, 2020a. Interscambio commerciale con l'estero della Provincia di Bergamo nel 4° trimestre 2019. accessed December 2020 from https://www.bg.camcom.it/sites/default/files/contenuto_redazione/rapporti/interscambio_commerciale_con_l_estero/2019-4t-import-export.pdf.
Camera di Commercio di Bergamo, 2020b. Osservatorio sulle imprese 4° trimestre 2019. accessed December 2020 from https://www.bg.camcom.it/sites/default/files/contenuto_redazione/rapporti/osservatorio_sulle_imprese/20194t-osservatorio-sulle-imprese.pdf.

Camera di Commercio di Milano, Monza e Lodi, 2020. Report import-export. IV trimestre 2019. accessed December 2020 from https://www.milomb.camcom.it/import-export.

Casti, E., 2015. Reflexive Cartography. A New Perspective on Mapping. Elsevier, Amsterdam.

Casti, E., 2020. Geografia a 'vele spiegate': analisi territoriale e mapping riflessivo sul COVID-19 in Italia. Documenti Geografici 1, 61–83.

Casti, E., Adobati, F. (Eds.), 2020a. Mapping riflessivo sul contagio del Covid-19. Dalla localizzazione del fenomeno all'importanza della sua dimensione territoriale. 2° Rapporto di Ricerca, L'evoluzione del contagio in relazione ai territori. CST, Università degli Studi di Bergamo, Bergamo. accessed December 2020 from https://medium.com/cst-diathesislab/2-rapporto-di-ricerca-e566a93431d3.

Casti, E., Adobati, F. (Eds.), 2020b. Mapping riflessivo sul contagio del Covid-19. Dalla localizzazione del fenomeno all'importanza della sua dimensione territoriale. 3° Rapporto di Ricerca, Le tre Italie. Fragilità dell'abitare mobile e urbanizzato. CST, Università degli studi di Bergamo, Bergamo. accessed December 2020 from https://cst.unibg.it/sites/cen06/files/3deg_rapporto_0.pdf.

Lavezzo, E., Franchin, E., Ciavarella, C., et al., 2020. Suppression of a SARS-CoV-2 outbreak in the Italian municipality of Vo. Nature 584, 425–429. https://doi.org/10.1038/s41586-020-2488-1.

Turri, E., 2000. La megalopoli padana. Marsilio, Venice.

Further reading

Lévy, J., 2008. Un évènement géographique. In: Lévy, J. (Ed.), L'invention du monde. Une géographie de la mondialisation. Presses de Sciences Po, Paris, pp. 11–16.

Lussault, M., 2019. Iper-luoghi. La nuova geografia della mondializzazione. Franco Angeli, Milan.

Pedersen, P.M., Ruihley, B.J., Li, B., 2020. Sport and the Pandemic. Perspectives on Covid-19's Impact on the Sport Industry. Routledge, London.

5.2

Nursing and Residential Care Facilities (RSA) and contagion-related fragilities in Italy

Marta Rodeschini

5.2.1 The role of nursing and residential care facilities during the Covid-19 epidemic

In the course of what has become a standard appointment with the World Health Organization in recent months, on 23 April 2020 the WHO regional director for Europe, Dr. Hans Henri P. Kluge, pointed out that nearly 50% global Covid-19 cases have so far occurred in the European region—over 1.2 million—and that over 110,000 people have lost their lives. He also lamented the severe strain placed on nursing and residential care facilities by the spread of the infection.[a] The WHO Director-General for Europe stated that according to estimates from member countries, up to half of Covid-19 fatalities occurred in long-term nursing and residential care facilities. Concerns are raised over operational

[a] April 23, 2020, Press conference, Dr. Hans Henri P. Kluge, WHO Regional Director for Europe: https://www.euro.who.int/en/health-topics/health-emergencies/coronavirus-covid-19/statements/statement-invest-in-the-overlooked-and-unsung-build-sustainable-people-center-long-term-care-in-the-wake-of-covid-19. Data on the evolution of the Covid-19 pandemic are available on health organizations websites: https://qap.ecdc.europa.eu/public/extensions/COVID-19/COVID-19.html; https://who.maps.arcgis.com/apps/opsdashboard/index.html#/ead3c6475654481ca51c248d52ab9c61.

procedures in these care facilities since their setup and care provisions seem to have favored the spread of the virus.[b]

A study by the International Long-Term Care Policy Network supports the statements of WHO representatives on the situation within Long-Term Care (LTC) residences[c] (Comas-Herrera et al., 2020). The main results of this research underline the difficulty of obtaining data for each country and discuss questionable data publication practices which de facto prevent researchers from accessing all figures pertaining to Covid-19 contagion and deaths, including data for LTC residents. Furthermore, data that cover provisions relating to Covid-19 deaths cannot be easily cross-referenced due to the range of different approaches whereby fatalities are recorded.[d]

With regard to Italy, the Istituto Superiore di Sanità (Higher Health Institute)—in collaboration with the National Watchdog for the rights of persons detained or deprived of personal liberty—set out to address this information gap and, starting from 24 March 2020, launched a targeted investigation into Covid-19 contagion in Nursing and Residential Care Facilities (RSAs) via a questionnaire that was distributed to all RSAs. A total of 1356 facilities, equal to 41.3% of those contacted, responded to the questionnaire, and reported data for the period from February 1 to April 30, 2020. The survey aimed to identify the main features of each facility, that is the makeup of the health workforce and the experts involved in the RSA; the number of beds and the number of residents; as well as the difficulties encountered in

---

[b] Attention is currently shifted to ways in which this sector of the health care system may be improved in previously neglected European regions. Key guidelines include (1) empowering health workers (who are often underpaid and unprotected); (2) changing the way health workers operate in long-term care facilities, with a view to striking a balance between the needs of residents and their families and the safe and secure management of services; (3) implementing a system that prioritizes people's needs, with the goal to fulfill their needs while preserving their dignity. In conclusion, WHO aims to ensure that services in long-term care facilities are safe and supportive, investing in staff training so that every person living in care systems may receive an adequate degree of attention and physical and psychological care.

[c] The International Long-term Care Policy Network was established in September 2010 and is based at the London School of Economics and Political Science (LSE), United Kingdom. It is a network of researchers, policy makers and stakeholders, who aim to promote a global exchange of research and knowledge on LTC policy.

In the spring of 2020, research at the International Long-term Care Policy Network was directed at identifying resources for supporting LTCs in the Covid-19 epidemic. More information may be found here: https://ltccovid.org/.

[d] Quantification of Covid-19 deaths follows three main approaches: (i) deaths of people who test positive for Covid-19 (either before or after their death); (ii) deaths of people who exhibit Covid-19 symptoms (based on symptoms); (iii) excess deaths (a comparison of the total number of deaths with those in the same weeks over previous years). Another issue is whether such data record the deaths of all LTC patients, who may have been transferred to hospitals as the disease worsened, or solely deaths that occurred within the care facilities themselves (Comas-Herrera et al., 2020, p. 4–5). Further information on global data collection on nursing homes deaths may be retrieved here: https://ltccovid.org/2020/04/12/mortality-associated-with-covid-19-outbreaks-in-care-homes-early-international-evidence/.

dealing professionally with the Covid-19 emergency.[e] As for Covid-19 deaths in Italian RSAs, out of a total of 9154 subjects who died, 680 had tested positive to a Covid-19 swab and 3092 had presented flu-like symptoms. Therefore, 7.4% of the total deaths tested positive for SARS-CoV-2 infection, while 33.8% involved residents with flu-like symptoms who were not given a swab-test.

A further study of the International Long-Term Care Policy Network on the Italian case (Berloto et al., 2020) outlines the multiple factors that contributed to Covid-19 spread in RSAs in Italy and exposed the weaknesses of the management system, such as the absence of unified planning in the design and development of a viable care model, hampered instead by a series of legislative procedures which de facto made the system hard to manage and control. The study points out that, even before the Covid-19 crisis, the Italian social and health care sector showed serious shortcomings, due to the high level of complexity and fragmentation in terms of skills and resources between institutional and non-institutional actors. And in the absence of a unified responsible institution, governance becomes inevitably difficult and cumbersome. It also laments the fact that such issues should not have been placed at the center of the political debate and ascribes this major shortcoming to the absence of a theoretical model for elderly care conceived and developed in unified terms. In short, the existence of nursing homes with different statutes and the numerous legislative interventions implemented over time have actually contributed to making the system difficult to manage.

The Covid-19 epidemic exacerbated the flaws of this management model. The International Long-Term Care Policy Network analytical report highlights three orders of issues: (i) poor and cumbersome management lines in the sector; (ii) delayed supply of personal protective equipment (PPE) to doctors and health workers in nursing homes; (iii) inability to control the spread of Covid-19 in RSAs.

This intricate administrative set-up for nursing homes comes clearly to the fore if we look closely at how care is provided locally and how the inconsistent distribution of RSAs throughout Italy and their overall number relative the elderly resident population ultimately engender serious shortcomings in the provision of an integrated and responsive health service.

The research quoted above indicates that the Covid-19 epidemic has exacerbated pre-existing weaknesses in the elderly care system. The overall picture for Italy is, however, far from homogeneous. Regional variation between North and South exists, based on locally prevailing cultural models. A much larger number of RSAs are located in the north than in the south, especially around the metropolitan area of Milan, where RSAs are numerous and large.[f] Conversely, the central and southern part of the Peninsula has comparatively fewer residential care facilities, despite a substantial presence of elderly people. This data may of course be interpreted in two ways: either as a shortcoming or, as evidence of a cultural model

---

[e]These include the lack of Personal Protective Equipment, the scarcity of information received about infection containment procedures, the shortage of drugs, the absence of health personnel and the difficulty in transferring residents affected by Covid-19 to hospitals, which in turn entails the difficulty of setting up adequate isolation procedures for residents affected by Covid-19 and the unavailability of swab-testing (Ancidoni et al., 2020, p. 19).

[f]An in-depth treatment of RSAs in the Lombardy region may be found in Chapter 5.4 of this volume.

that offers an alternative form of elderly care, provided at home directly by the family group.[8] In Southern Italian regions, in fact, the health care system relies on an in-family caregiver, i.e., on help that the family gives to a relative who is not self-sufficient, an issue we will address shortly.

In the same section, reflexive mapping will be used to highlight regional differences by cross-referencing them with this cultural model. Such mapping suggests the need for a careful reassessment of RSAs' organization (Azara et al., 2013, p. 187).

5.2.2 Elderly care and nursing and residential care facilities (RSAs) in Italy

Elderly healthcare in Italy is provided by the health services put in place by the National Health System (SSN), and is broken down into hospital care, integrated home care (ADI), nursing and residential care (RSA) and semi-residential care (day care centers). In addition to these welfare policies, there exist forms of financial support which cover old-age pensions and disability or poverty pensions: these provide complementary monetary aid to support people who are not self-sufficient (disability living allowance, checks, etc.) (Azara et al., 2013, p. 185). The National Health System provides a range of services in this sense, in accordance with policies and guidelines issued by the Government and acting in concert mainly with the Ministry of Social Policies, but also with the Ministry of Labor and the Ministry of Health.[h] At a local scale, Italian regions are entrusted with translating ministerial guidelines into planning guidelines within their territories. They must also lay out their working agenda and outline the service network to be offered. Implementation of actual services and interventions is subsequently entrusted to local Municipalities as regards social care and to the Local Health Authorities (ASLs)[i] as regards healthcare; the supply of services is also ensured by ASLs, by municipalities or by private contractors (Rotolo, 2014, p. 94).[j]

Home services are also mainly provided by the two public bodies mentioned bodies, namely ASLs and Municipalities. In collaboration with Municipalities, ASLs set up Integrated Home Assistance (ADI), whereby non self-sufficient persons are assisted by nurses or other health personnel at home, thus avoiding the need for admissions or shortening

---

[8]With the onset of disease, most of the responsibility for health care falls primarily on the relatives, either spouses or children, who therefore need to introduce profound changes in their daily lives. Family members cease to be just husband and wife, or son and daughter: they also become "caregivers" (Eifert et al., 2015, p. 357).

[h]A core level is covered by INPS (National Social Security Institute) which, in agreement with the Ministry of Social Policies, has the task of defining the reference framework for the economic part, such as the allocation of disability living allowances and other services intended for non-self-sufficient persons.

[i]A Local Health Authority (ASL) is a body of the Italian public administration, responsible for providing health services. They may be named differently throughout Italian regions, but refer to the same functions. ASLs comprise various departments, most notably territorial health services and hospitals, and may include a range of sectors, such as: consultancy, prevention department, pathological addiction services, home treatment and residential care services.

[j]Detailed information on the allocation of healthcare tasks and responsibilities between the Italian State and individual Italian regions may be found in Chapter 5.3 of this volume.

hospitalization times.[k] Whenever home care is not possible or deemed inappropriate, a number of targeted residential and semi-residential care services are offered to non-self-sufficient persons. In the first case, we have hotel-like facilities, which either temporarily or permanently host non-self-sufficient persons. In the case of semi-residential care, care is ensured by day-type facilities (for example, day care centers) which support the elderly. The latter service is intended for people who are partially self-sufficient or exhibit early signs of cognitive decline (Perobelli and Notarnicola, 2018, p. 21).

The planning of services provided by this care sector, as a whole, must take into account the fact that, in recent decades, Italy's demographic setup has undergone significant changes (ISTAT, 2020, p. 255). In this regard, demographic analyses conducted before the onset of Covid-19 suggested that the Italian population was in serious decline and could be expected to fall to 59.3 million by 2040 and to 53.8 million by 2065,[l] against the current figure of over 60 million[m] residents. This process is closely tied to population aging, since current birth rates are too low to make up for deaths. Life expectancy is bound to increase yearly, so much that by 2065 it is projected to reach over 86 years for men and 90 years for women.[n] Furthermore, the quality of life indicators show that the health conditions in Italy have improved and, consequently, the average life expectancy (longevity) in good health for a large number of elderly people has increased (ISTAT, 2020, p. 122). This demographic process suggests that an elderly population will be putting forward new welfare needs, mostly to do with health care.

This change of perspective is particularly evident at present, if we turn to consider the residential healthcare system. Residential healthcare over the last two decades has gradually changed: originally, it was established as a service mostly aimed at hotel use. In time, however, it has taken on a purely healthcare function (Giunco et al., 2013, p. 14). To this should be added that, over the last century, we have witnessed the insurgence of the phenomenon named "epidemiological transition," whereby infectious and parasitic diseases that were prevalent in the first half of the last century nearly disappeared to be replaced by chronic-degenerative diseases.[o] These require patients to be assisted in specialized facilities separated

---

[k] Various types of ADI exist, which differ according to the intensity of the care required and the set of health professionals involved. Local Municipalities also contribute to supplementing home care, by setting up their own Home Care Service, which may have a range of different regional denominations such as SAD, AD or otherwise, and is characterized by targeted care with distinctive social relevance.

[l] Population data for this research dates back to 1 January 2018 (ISTAT, 2020, p. 255).

[m] This forecast was based on a median scenario, which ISTAT claim should be related to the "age that divides a population into two numerically equal groups; that is one group with a population younger than the one identified, and a second group with an older population" (ISTAT, 2020, p. 277). From a statistical viewpoint, this claim discloses the criterion or value adopted in this forecast.

[n] Overall, the term "society aging" defines all the social changes that affect a population due to a combination of phenomena: (a) birth rate decrease, (b) infant mortality increase and (c) life expectancy increase at birth (Fara and D'Alessandro, 2015, p. 21).

[o] "Chronic-degenerative" describes clinical pictures that do not require hospitalization, but call for assistance either from a home-care network or from specifically designed facilities (Azara et al., 2013, p. 183).

from the hospitals, which must be relieved from excessive strain and allowed to concentrate mainly on acute conditions.[P]

For the sake of completeness, it should be remembered that assistance to non-self-sufficient elderly people does not rely exclusively on formal public and private services. Rather, it receives strong support from what is defined as "informal" care, provided free of charge by family members, friends, acquaintances, and/or volunteers (Azara et al., 2013, p. 186). These people, who help a family member in daily activities, care management, general support or other form of assistance, are also called family caregivers[q] (Montemurro and Petrella, 2016, p. 52). Informal family care in Italy is recognized as the pivot of the welfare system, a marker of responsibility and awareness of the fact that families provide crucial care, the costs of which fall mostly on women (Montemurro and Petrella, 2016, p. 14). Within Italian families the burden of care falls mostly on women. As evidence of this, it was noted that the drop in family care that has occurred in recent years is linked to the growing participation of women in the labor market and to the introduction of new social security legislation, which contribute to a progressive decline of family care for non-self-sufficient elderly people.[r] Regional differences in this regard are accentuated relative to dissimilar percentages of female labor participation: ISTAT data confirm that 33.2% of women between the ages of 15 and 64 are employed in Southern Italy. The percentage is nearly double in the rest of Italy, where it reaches 59.5%.[s]

Fig. 5.4 confirms this state of affairs indirectly. By surveying the distribution of nursing facilities throughout the national territory, it highlights their limited presence in the Southern part of the peninsula, where family caregivers take up much of the burden of care. The map records an unbalanced distribution of RSAs in the north and the south, with significantly fewer facilities in central-southern regions, which however have consistently larger numbers of elderly residents. The regions of Campania and Apulia provide two telling instances in this respect.

In our map, data relating to the number of Nursing and Residential Care Facility in each region were also cross-referenced to the number of elderly residents. This is a reflexive map,[t] which uses information cross-referencing to chart the complex layering of the phenomenon it represents. In this instance, reflexive mapping makes it possible to extrapolate relevant social

---

[P] Acute care, that sets in at the stage of maximum severity and intensity of a disease, has needs diametrically opposed to hospitalization requirements for chronic-degenerative diseases. In fact, acuity calls for intensive and targeted assistance and relies on the shortest possible hospitalization.

[q] Family *caregivers* may be backed by family assistants, i.e. workers directly hired by the family or by the non-self-sufficient elderly person (Di Rosa et al., 2015, p. 38).

[r] As a result of this, in fact, individual family members have become less willing to fully take up family care, which in turn has entailed a progressive reduction in the number of children and family members (Montemurro and Petrella, 2016, p. 15).

[s] Data are sourced from ISTAT analyses, available online: http://dati.istat.it/Index.aspx?DataSetCode=DCCV_TAXOCCU1. Southern Italian regions include Abruzzo, Molise, Campania, Apulia, Basilicata, Calabria and the islands of Sicily and Sardinia.

[t] By reflexivity we mean "an analytical perspective aimed at achieving practical goals by acting upon researchers. Researchers are made to reflect upon themselves and the actions they are doing or have done; actions which in turn sketch the scenario in which they are to operate" (Casti, 2015, p. 260).

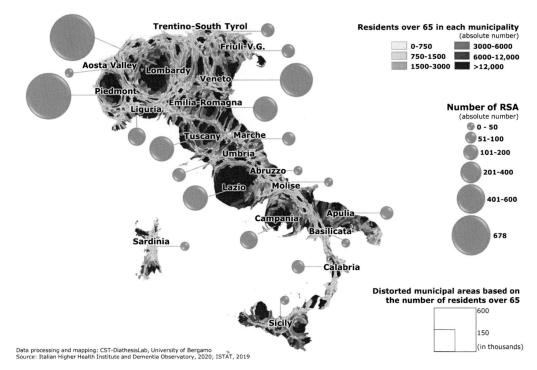

FIG. 5.4 RSA distribution across the Italian regions in relation to elderly inhabitants. *Note: The number of RSAs by region is provided in the National Survey Report on Covid-19 infection in residential and social and health structures, issued by the Higher Health Institute (Istituto Superiore di Sanità). According to ISTAT, data on elderly residents refers to January 1, 2019.*

data, since the Covid-19 epidemic impacts a specific population sector, elderly persons who are most easily attacked by the virus.[u]

The map's semantics uses multiple visualization systems: the map ground does not reflect the scaling of municipalities but is distorted anamorphically to render the number of elderly residents (i.e., residents over 65). Anamorphic visualization effectively shows that such residents mostly inhabit urban municipalities and most notably metropolitan ones, where their presence is high. Color-coding makes it possible to such presence at a glance: color intensity uses a six-gradient range: municipalities with fewer than 750 residents aged over 65 use a faint light-brown shade, while large municipalities, which number more than 12,000 elderly, are shown in solid brown. Color gradients also indicate that the elderly are found mainly in

[u]The theory of cartographic semiosis, which is adopted in this study, highlights how maps contain multiple information levels, given by their hypertextual language, and may be conceived as a blend of communication systems that bring out specific territorial features. This theory takes territory as the result of a process whereby natural space incorporates anthropological values. Two communicative levels apply here: the connotative and the denotative. The latter refers to the most superficial level of communication, while the connotative level refers to socially accumulated knowledge that may be logically arranged only when laid out on a map through certain semantic methods (Casti, 2000, pp. 17–34).

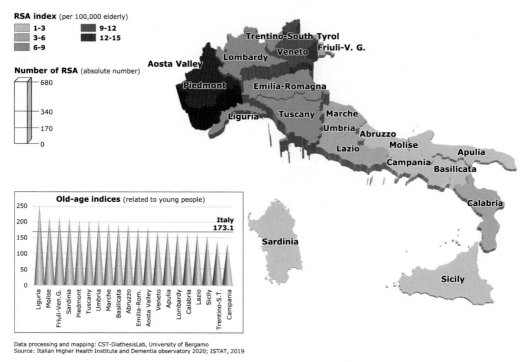

FIG. 5.5   Distribution of RSA and old-age indices by region.

urban areas, especially in metropolitan clusters (such as Milan, Rome, Naples, Turin, Palermo) and in large cities. Circles were superimposed onto the map ground. Their varying sizes represent the number of RSAs for each Italian region. As mentioned above, Northern regions have the highest number of RSAs, namely Lombardy and Piedmont, followed by Veneto, Lazio, Tuscany, and Emilia-Romagna.

By changing the map's graphic semantics, we can process data in an alternative model, as may be seen in the Fig. 5.5 map, which yields a different interpretive set. This last map does not refer to the absolute number of elderly people, but to an old age index, shown in the histogram on the left. RSAs, on the other hand, are represented both in their absolute number—which the map shows by extrusion of regional areas, whose thickness thus reflects their numbers—and on the basis of the index, that is the ratio between the number of RSAs per 100,000 elderly inhabitants per region. The combination of these infographic tools informs that Northern Italy has a higher percentage than Southern Italy. However, once we get into details, the Northern regions have considerable dissimilarities: the region with the highest numbers and index of health care facilities is Piedmont, even though its old age index (in the histogram) ranks fifth in Italy. Veneto trails behind with a slightly lower number of RSAs which, however, must be referenced to a lower old age index, for which Veneto ranks thirteenth. Finally, despite a substantial absolute number of RSAs with their own medium-low index, Lombardy hosts a population with a medium-high old age index, for which Lombardy ranks seventh on the national list. A mention should also be made of Liguria, for even though in terms of elderly population it has the highest old-age index, both its absolute number of RSAs and its RSA index are relatively low.

The different thickness of central-southern regions on the map indicates that they have a low number of RSAs both in absolute terms and on the basis of the index. It should be remembered, however, that the old-age index for these regions is also more contained. The general picture that emerges from this shows a mismatch between the number of elderly people and the presence of care facilities. This has to do with regional health policies which, albeit in compliance with top-down government guidelines, can effectively determine the quality or type of care.[v]

If we focus our analysis solely on the histogram that charts the old-age index for each Italian region,[w] we conclude that the region with the highest old-age index is Liguria, while the one with the lowest index is Campania. Also, there seems to be no match between the seniority index and the RSA index for southern regions. An instance of this is the region of Molise, which appears to have a very high old age index, above the national average, but has a very low RSA index relative to the elderly inhabitants. Other regions such as Sardinia or Basilicata have similar features.

5.2.3 Conclusions

During the Covid-19 epidemic, RSAs became sites of viral implosion. Mortality rate in RSAs was high, especially in territories where the contagion index was high, i.e., Northern Italy, even though it must be noted that the severity of the epidemic wave in the rest of the country was far more contained.[x] Ultimately, regional differences between Northern and Southern Italy reveal a discrepancy in the provision of health care for non-self-sufficient persons, which in the course of the Covid-19 epidemic has grown worse and has led to lethal outcomes.

Our analysis has also pointed out that demographic and social changes are progressively modifying basic family structures and, along with these, also the network of informal help provided by family members, relatives, friends or neighbors. Informal assistance of this kind has become brittle and seems now inadequate to care for the needs of the elderly (Montemurro and Petrella, 2016, p. 4). In the past, the elderly could live in their own homes thanks to such informal care. However, in the last decade social changes in society, as well as a progressive increase in life expectancy and greater female employment have made this solution no longer viable. Informal care survives only in a few inner areas of Southern Italy, where

[v] On current legislation and its functioning, see Chapter 6 in this volume.

[w] According to ISTAT, old-age index is the ratio between the population aged 65 and over and the population aged 0–14, multiplied by 100 (http://demo.istat.it/altridati/indicatori/index.html). Data used in this processing is provided by ISTAT via an analysis of demographic indicators. For further details see: https://www.istat.it/it/archive/238447.

[x] The second wave of contagion, however, has affected the entire Italian peninsula, so that the central-southern regions have reached the same levels of infection as the northern ones. See the dashboard of the Italian Ministry of Health: http://opendatadpc.maps.arcgis.com/apps/opsdashboard/index.html#/b0c68bce2cce478eaac82fe38d4138b1.

female employment is low and formal care services lack. While negative in themselves, these data may point to a prospective care model that is arguably more resilient in the face of pathologies than the one offered by RSAs. As a matter of fact, the current European approach to elderly care is going precisely towards enhancing home care and non-institutionalized assistance; towards boosting community services and bringing about an overhaul of public services that aims to promote home care (Montemurro and Petrella, 2016, p. 9).

The epidemic crisis in Italy has exposed structural deficiencies in the elderly care system: population aging, and the possibility of novel epidemics, demand urgent attention. The approach to be taken is in fact established and now clearer than ever: we need to pursue a new care model that addresses at the same time family needs and gender values, while making it possible for the elderly to live the rest of their lives peacefully and safely.

References

Ancidoni, A., et al., 2020. Survey nazionale sul contagio Covid-19 nelle strutture residenziali e sociosanitarie. Istituto Superiore di Sanità, Roma. accessed October 2020: https://www.epicentro.iss.it/coronavirus/sars-cov-2-survey-rsa.

Azara, A., et al., 2013. L'assistenza sanitaria per gli anziani tra ospedale e territorio. In: Annali di igiene. Medicina preventiva e di comunità. 25. Società editrice universo, Roma, pp. 183–188. 3.

Berloto, S., Notarnicola, N., Perobelli, E., Rotolo, A., 2020. Report on Covid-19 and Long-Term Care in Italy: Lessons Learned from an Absent Crisis Management. LTC Responses to Covid-19, International Long-Term Care Policy Network. accessed October 2020 from: https://ltccovid.org/2020/04/10/report-on-covid-19-and-long-term-care-in-italy-lessons-learned-from-an-absent-crisis-management/.

Casti, E., 2000. Reality as Representation. The Semiotics of Cartography and the Generation of Meaning. Edizioni Sestante, Bergamo.

Casti, E., 2015. Reflexive Cartography. A New Perspective on Mapping. Elsevier, Amsterdam.

Comas-Herrera, A., et al., 2020. Mortality Associated with Covid-19 Outbreaks in Care Homes: Early International Evidence. LTC Responses to Covid-19, International Long-term Care Policy Network. accessed October 2020 from: https://ltccovid.org/2020/04/12/mortality-associated-with-covid-19-outbreaks-in-care-homes-early-international-evidence/.

Di Rosa, M., Barbabella, F., Poli, A., Balducci, F., 2015. L'altra bussola: le strategie di sostegno familiare e privato. In: L'assistenza agli anziani non autosufficienti in Italia, Un futuro da ricostruire, 5° Rapporto. Maggioli, Santarcangelo di Romagna, pp. 35–54.

Eifert, E.K., Adams, R., Dudley, W., Perko, M., 2015. Family caregiver identity: a literature review. Am. J. Health Educ. 46 (6), 357–367. https://doi.org/10.1080/19325037.2015.1099482.

Fara, G.M., D'Alessandro, D., 2015. L'invecchiamento della popolazione: riflessi sulla soddisfazione delle esigenze socio-assistenziali. Techne. J. Technol. Architect. Environ. Architect. Health Educ. 9, 21–26.

Giunco, F., Predazzi, M., Costa, G., 2013. Verso nuovi modelli di residenzialità. Il progetto Abitare Leggero. I luoghi della cura. Vol. XI CIC edizioni internazionali, Roma, pp. 13–20. 4.

ISTAT, 2020. Rapporto annuale 2020. Roma. accessed October 2020: https://www.istat.it/it/archivio/244848.

Montemurro, F., Petrella, A., 2016. Le politiche per gli anziani non autosufficienti nelle regioni italiane. In: Morosini, I. (Ed.), Cgil Spi. Torino.

Perobelli, E., Notarnicola, E., 2018. Il settore Long Term Care: bisogno, servizi, utenti e risorse tra pubblico e privato. In: Fosti, G., Notarnicola, E. (Eds.), L'innovazione e il cambiamento nel settore delle Long Term Care. 1° Rapporto Osservatorio Long Term Care, Egea, Milano, pp. 19–48.

Rotolo, A., 2014. Italia. In: Fosti, G., Notarnicola, E. (Eds.), Il Welfare e la Long Term Care in Europa. Modelli istituzionali e percorsi degli utenti, Egea, Milano, pp. 93–114.

Further reading

ECDC Public Health Emergency Team, et al., 2020. High impact of Covid-19 in long-term care facilities, suggestion for monitoring in the EU/EEA, May 2020. Euro Surveill. 25 (22), 1–5. https://doi.org/10.2807/1560-7917.ES.2020.25.22.2000956.

European Centre for Disease Prevention and Control, 2020. Surveillance of Covid-19 at Long-Term Care Facilities in the EU/EEA. ECDC, Stockholm.

Gardner, W., States, D., Bagley, N., 2020. The coronavirus and the risks to the elderly in long-term care. J. Aging Soc. Policy 32 (4–5), 310–315. https://doi.org/10.1080/08959420.2020.1750543.

Spasova, S., Baeten, R., Coster, S., Ghailani, D., Peña-Casas, R., Vanhercke, B., 2018. Challenges in long-term care in Europe. In: A Study of National Policies. European Social Policy Network (ESPN), European Commission, Brussels.

5.3

The Italian health care system and swab testing

Marta Rodeschini

5.3.1 The healthcare system in Italy

At a national level, differences in the management of the Covid-19 epidemic are determined by the setup of the Italian healthcare system, which covers a complex layering of functions, activities and care services managed and provided independently by individual Regions. It is a public system that guarantees health care through the provision of state and regional competences, in accordance with Article 32 of the Constitution of the Italian Republic[a] which establishes the principle of subsidiarity.[b] By virtue of this, the National Health Service (SSN) deploys its competences between the State and the Regions. More specifically: at a central level, the state is responsible for ensuring all citizens are entitled to health care via a solid system of guarantees, encoded as Essential Levels of Assistance (LEA). At the regional level, on the other hand, government bodies bear direct responsibility for implementation and expenditure, so as to achieve the country's health objectives. LEAs cover performances and services that the SSN is required to provide to all citizens. A committee is responsible for monitoring LEAs annually, to verify that all regions maintain

---

[a]It states that: "The Republic safeguards health as a fundamental right of the individual and as a collective interest, and guarantees free medical care to the indigent. No one may be obliged to undergo any health treatment except under the provisions of the law. The law may not under any circumstances violate the limits imposed by respect for the human person."

[b]This principle is based on the assumption that if a lower institution is able to perform a task well, the higher body should not intervene, but may possibly support local action. This principle is defined in art. 5, paragraph 3, of the Treaty on European Union (TEU) and in Protocol no. 2 on the application of the principles of subsidiarity and proportionality of the consolidated version of the Treaty on European Union. For more information see: https://www.europarl.europa.eu/factsheets/en/sheet/7/il-principio-di-sussidiarieta.

the agreed standards.[c] It is therefore the regions that have exclusive competence in the regulation and organization of services and activities intended for the protection of health and of the funding of Local Health Authorities[d] (ASL) and hospitals. The same applies to management control and quality control for all health services, in compliance with the general guidelines established by the laws of the State.[e] The national health service therefore does not consist of a single administrative structure. Rather, it is a set of institutions and bodies which contribute to achieving the objectives of protecting the health of citizens. The Ministry of Health coordinates the national health plan, thanks to the constitutionally enshrined competences of the Regions.[f]

In compliance with the public health service thus enshrined, on January 31, 2020 the Italian Council of Ministers approved a state of national emergency.[g] A state of emergency[h] in Italy may be decreed whenever critical junctures arise that may affect human health. Such emergency endows the government with "extraordinary" or "special" powers. Declaring a state of emergency therefore makes it possible to implement extraordinary interventions, with ordinances in derogation of normal legal provisions. For example, it is permissible to plan swift medical aid responses; to issue awareness-raising guidelines on behaviors to be adopted in the event of risk; to activate rescue procedures provided for in municipal, provincial and regional plans.

As for the state of emergency that was announced following the Covid-19 epidemic, the Government, in agreement with the Ministry of Health, took on the task of coordinating responses to the epidemic, and specifically the issue of swab testing. On February 25, 2020, the Ministry issued Ministerial Circular (Memorandum) no. 0005889, with the resolution that:

[c] For further information regarding the monitoring system, the standing committee for verifying the provision of the Essential Levels of Assistance and the obligations to be observed for maintaining LEA requirements, you may consult the page of the Italian Ministry of Health at: http://www.salute.gov.it/portale/lea/menuContenutoLea.jsp?lingua=italiano&area=Lea&menu=monitoraggioLea.

[d] ASLs may be named differently throughout Italian regions, but refer to the same functions. Their key function rests in the organization and planning of the care system itself, as well as in the provision of services.

[e] As mentioned above, the deployment of healthcare services is allocated on the basis of different levels of responsibility and governance. The National Health Service consists of several bodies, whose main task is to help achieve the objectives of protecting the health of citizens. For more information see: https://www.salute.gov.it/portale/ministro/p4_5_5_1.jsp?lingua=italiano&label=org&menu=organizzazione.

[f] Title V of the Italian Constitution legislates on health matters, and may be consulted here: http://www.governo.it/it/costituzione-italiana/parte-seconda-ordinamento-della-repubblica/titolo-v-le-regionile-province-e-i. Article 117 of the Italian Constitution also establishes that the State retains exclusive legislative competence in a series of specifically listed matters. Paragraph 3 of the same article decrees that the Regions can legislate in matters of shared competence, in compliance with the fundamental principles defined by the State.

[g] The resolution of the Council of Ministers of January 31, 2020, entitled "Declaration of the state of emergency as a consequence of the health risk associated with the onset of pathologies deriving from transmissible viral agents" may be accessed in a dedicated section of the Italian Gazzetta Ufficiale: https://www.gazzettaufficiale.it/eli/id/2020/02/01/20A00737/sg.

[h] To cope with national-scale emergencies that require immediate intervention and must be dealt with extraordinary means and powers, the Council of Ministers approves a state of emergency declaration, upon proposal of the Prime Minister, once agreement with the region concerned has been reached.

"Swabs should be performed only for symptomatic cases of ILI (Influenza-Like Illness) and SARI (Severe Acute Respiratory Infections), as well as for suspected cases of Covid-19 according to definition given in Annex 1 of the aforementioned memo. In the absence of symptoms, therefore, the test does not appear to have adequate scientific grounds, since it falls short of providing indicative information for clinical purposes consistent with the definition of 'case'.

It should be remembered that the procedure for definitive confirmation of a given case is entrusted to the Higher Health Institute (Istituto Superiore di Sanità ISS). Therefore, a case may not be defined as confirmed without the aforementioned validation by an ISS laboratory. For this reason, the need to always and promptly send suitable samples to the aforementioned ISS must be underlined."[i]

This means that since February 25, Covid-19 swabs in Italy have been carried out solely on subjects who presented symptoms. And only those who tested positive were then classified as "suspect" cases.[j]

Fig. 5.6 charts the tests' space–time quantification and their outcomes from March 23 to May 20, 2020. The map ground indicates the absolute number of the total people infected by region as of May 20, 2020 and highlights the profound difference between the north and the south of Italy: in the north the number of infections reaches very high peaks; in the center infections are reduced but remain severe; in the south, instead, infections are limited. The graphs instead show the number of swabs carried out by region and chart their outcome (either positive or negative) for the seven dates[k] our analysis has focused on to monitor the infection's progress at regular intervals. In all regions, there emerges a progression of positive outcomes over time, albeit in a limited ratio relative to the number of swabs performed. The regions that carried out the most swab tests are Lombardy (607,863 as of May 20), followed by Veneto (536,798), then by Emilia-Romagna (274,362) and Piedmont (264,624). Lazio (217,849) and Tuscany (214,299) are the regions of central Italy with the

[i]To look at the Ministerial Circular or Memorandum in full see: https://www.gazzettaufficiale.it/eli/id/2020/02/26/20A01300/sg.

[j]There are two tests for Covid-19 detection using pharyngeal "swabs": the first, a molecular test, analyzes the amplification of viral genes in the course of infection. This type of tests has been carried out regularly since the beginning of the epidemic in highly specialized labs formally endorsed by the Italian health authorities. An average of 2 to 6 h is needed for a molecular test report. The second type of test is an antigen test, which has been in use in Italy since September 24, 2020. It is a rapid test based on the search for viral proteins (antigens). Response times in this case are very short (about 15 min). In addition, there are two other types of serological screening: an immunological test, and a rapid serological test, both of which look for specific antibodies in the blood in response to infection. The former establishes the presence, type and quantity of serum antibodies and is mostly carried out in specialized labs. The latter relies on a simplified procedure which only provides a simplified qualitative response. Both are in fact relatively easy to administer, have an average response time of about 15 min and may also be performed outside labs. However, since the reliability of this type of test is highly variable, the World Health Organization does not recommend it for the detection of antibodies in patient care (Ministero della Salute et al., 2020, p. 8–9). Furthermore, Ministry of Health Memo of September 29, 2020, protocol number 31400 underlines that molecular tests remain the safest ones for a SARS-CoV-2 diagnosis. These tests are collected by the regions and passed on to the National Institute of Health.

[k]The following seven dates are analysis addressed: March 23, March 30, April 6, April 16, April 28, May 14 and May 20, 2020.

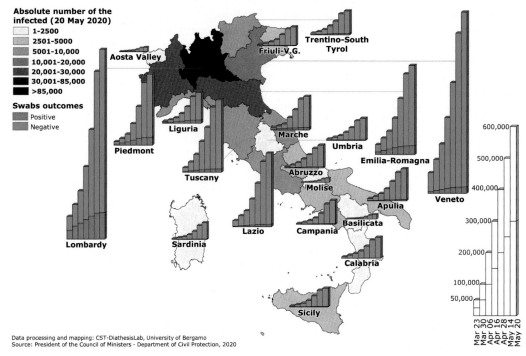

FIG. 5.6 Evolution of regional number of Covid-19 infected in relation to the swabs carried out. *Note: The Ministry of Health provides data on swab tests carried out by region. It should be pointed out that the number of tests includes all swabs administered: it is not infrequent for a subject to have been tested for Covid-19 via nasopharyngeal swab.*

highest degree of swab-testing. However, if we take the data not in absolute terms, but as a percentage (Fig. 5.7) both as regards the testing and as regards results, it becomes obvious that in various regions, especially in Northern Italy, the percentage of positive swabs is high, to the point that it even exceeds a quarter of the total tests performed until mid-April. The regions that administered more swab tests to infected people are Lombardy, Piedmont, Emilia-Romagna, Liguria, and Marche. In these regions, especially in the first weeks of the epidemic, nearly 25% of the tests were positive. Aerograms for infection percentages are sized with respect to the total number of swab tests administered. Lombardy stands out as the region that performed the greatest number of swab-tests, but also as the region recording the highest number of infections. The map ground shows the incidence of Covid-19 infection as of May 20, that is to say the number of infected people calculated against the total population.[1]

Lack of consistency in the administration of swab tests throughout the nation suggests that even the dataset that charts the average age of infected persons, and the evolution of such

---

[1]Lombardy and Aosta Valley are the regions with the highest infection incidence, followed by the northern regions of Piedmont, Trentino-Alto Adige and Emilia-Romagna. Liguria and Marche are listed before Veneto, Friuli-Venezia Giulia and the other regions of central Italy. Conversely, southern Italy has a low incidence of contagion.

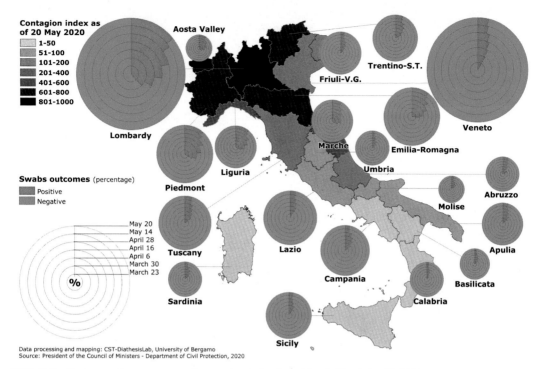

FIG. 5.7 Region swabs outcomes and contagion index from March 23 to May 20, 2020.

dataset over time, ought to be taken with caution. This is due to the fact that data are drawn from testing that was carried out without a representative sampling of the entire population, and based on a selection dependent on the contingency of the epidemic.[m] Accordingly, Fig. 5.8, which shows a breakdown of infected persons by age group as of March 23, indicates that Lombardy has the highest contagion rate in the age group between 70 and 79 years of age, just like Aosta Valley, Emilia-Romagna, Piedmont, and Liguria. In Veneto, Trentino-Alto Adige, Friuli-Venezia Giulia, and Tuscany, on the other hand, the most affected age band was that between 50 and 59 years of age. The map represents contagion data for March 23 and May 20, which makes it possible to chart the contagion's development. This map confirms the initial trend also over the following weeks: for nearly all regions, the age band with the highest contagion rate goes from 50 to 59 years of age, except for Lombardy, Aosta Valley, Piedmont, and Liguria, which show a prevalence of infection both in the 50 to 59 age range and in the higher ranges. Contagion in the 80 to 89 age range is in fact considerably higher than in other age groups in Lombardy, Piedmont, and Liguria. These data may be taken to suggest that there exist different contagion modes for different regions and that in some areas the elderly may be more affected while in others the adult population is. However, such data

[m]Sampling plays an essential role in epidemiological analyses, since the purpose of sampling is to extract a representative sample of features for the target population (Franco and Di Napoli, 2019, p. 171). Carrying out an epidemiological analysis on a representative sample of the population is a prerequisite for such analysis to produce reliable results.

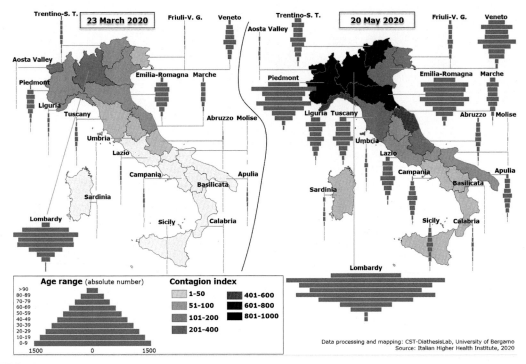

FIG. 5.8 Evolution of the infection by age range in Italy from March 23 to May 20, 2020.

depend on the population sample that was swab-tested, which varies from region to region, and thus yields different, non-comparable datasets which are influenced by external and non-homogeneous factors.[n]

To corroborate this hypothesis, research conducted by the epidemiological service of the Azienda per la Tutela della Salute (Health Protection Agency) in Bergamo classifies by age and sex the whole population of individuals who tested positive for Covid-19 during the observation period from February 17, 2020 to September 3, 2020. The study confirms that, in the case of the Bergamo area, swab-testing policies reached specific population groups, thus producing an oscillation in the records of the most affected age groups. As a result, the most infected age groups are found to change according to the phases of the epidemic. In the most acute epidemic phase, it was the subjects who requested hospital treatment that were almost exclusively tested, that is only a fraction of the fragile population in terms of chronic diseases, which on average correlate to age.[o]

[n]For the purposes of comparing Covid-19 impact across regions, it is therefore more reliable to analyze mortality in each region, which may provide a consistent overview of Covid-19 virulence in Italy. For an in-depth treatment of this issue see Chapter 2 of this volume.

[o]In the period of March–April, which had median ages even higher than those recorded in the acute phase, RSA residents were also subjected to massive swab-testing. In the weeks that followed, numerous serological screening programs were set up over the population of active working age, with subsequent swabs and the consequent effect of reducing the median age among those who swab-tested positive (Zucchi and Ciampichini, 2020).

5.3.2 The epicenter regions of the epidemic

In order to assess the claim whereby regional health systems have produced dissimilar monitoring strategies for Covid-19 infection, we will compare the regions of Lombardy and Veneto. Regional differences in the management of Covid-19, and therefore also in the administration of swab-tests, did depend on a range of factors, such as regional setup of health services, contagion's impact and virulence, regional ordinances, availability of medical equipment (such as personal protective equipment or suitable reagents for swab analysis).

In Veneto, for example, by initiative of the Director of the Department of Molecular Medicine at the University of Padua, Professor Andrea Crisanti, the region's health management was informed as early as January 20 that the University laboratory would be making a large purchase, in order to ensure a sufficient supply of reagents for analyzing nearly 500,000 swabs for SARS-CoV-2 identification in patients.[p]

The first cases attributable to Covid-19 in the Veneto region were detected on February 20 in Vo' Euganeo, a municipality with just over 3000 residents. The area was declared a "red zone" and placed under isolation on February 24. Between February 24 and March 2, 2020, Covid-19 cases in Veneto increased by 8.5 times, going from 32 to 271 (from 0.6/100,000 to 5/100,000). At that point, the Veneto health authorities identified hospitals and convalescent homes to be allocated for dealing with Covid-19 cases. They also doubled the region's intensive care capacity and ensured an adequate supply of ventilators. Non-Covid-19 patients were gradually moved from epidemic-allocated hospitals to smaller community facilities. In addition to strengthening patient care capacity, regional health authorities in Veneto developed and implemented public health measures: in coordination with local managers, they implemented a complex regional strategy, which included extensive contact tracing, rapid case testing and an extended contact network, supervised quarantine and isolation, minimization of contact between healthcare professionals and the public, IT tools for rapid communication on diagnosis and case management and for monitoring bed availability. All non-essential public health activities were swiftly suspended and a labor force of over 750 public health workers was mobilized across the region.[q]

As for Lombardy, the first case was identified on February 20 in Codogno,[r] a town of 15,000 inhabitants located near the city of Lodi. Codogno was placed under isolation by the national government on February 24. In the space of 7 days, from February 24 to March 2, the number

---

[p]This information is attested by multiple media sources and interviews with Prof. Crisanti. Among these: https://www.ilpost.it/2020/04/16/coronavirus-veneto-modello/; https://www.wired.it/scienza/medicina/2020/04/20/coronavirus-vo-euganeo/?refresh_ce=.

[q]With regional council decree no. 269 of March 2, 2020, the Veneto region reorganized healthcare activities in order to preserve intensive care beds. Also, on March 14 Veneto launched a specific Public Health Plan, which aimed to break all possible transmission chains of the Covid-19 virus by pinpointing all possible suspect, probable and confirmed cases and carrying out an in-depth epidemiological investigation with a view to identifying all possible contacts. To that end, it was decided to broaden swab-testing to include also occasional contacts, with the intent to isolate asymptomatic positives as well, and to consider the possibility of focusing on specific categories that might be particularly exposed (Veneto Region press release, March 14, 2020, available at: https://www.regione.veneto.it/article-detail?articleGroupId=10136&articleId=4377232).

[r]For an informed discussion of reticularity in viral spread and the localization of first outbreaks, see Chapter 5.1. in this volume.

of cases in Lombardy went from 166 to 1077 (from 1.6 to 12/100,000 inhabitants). The initial efforts of the regional health system focused on three primary objectives: (i) gathering data in order to understand epidemiological trends and to model intervention; (ii) boosting diagnostic capacity; (iii) promoting hospital care. Efforts were made to isolate and trace contacts and, in the meantime, the solid regional ICU—Intensive Care Unit network that was already in place was strengthened (Binkin et al., 2020, p. 6).

The Region issued guidelines for primary care physicians regarding diagnosis, testing and hospital referral.[s] As per the national directive of 25 February cited above, the tracing of infections in Lombardy was focused on persons with symptoms and the tracking of their contacts. Home-administered swab tests, care and diagnosis were hampered by the rapid outburst in the number of Covid-19 cases. For lack of other options, patients in Lombardy were sent to hospital, straining human resources and the availability of hospital beds. Wards entirely dedicated to Covid-19 were set up inside existing health facilities. Due to the virulence of contagion, adequate ward segregation proved impossible, especially for virus-positive patients whose conditions did not require intensive or sub-intensive care.[t] In fact, convalescence centers for those who did not need intensive care but required continuous monitoring were only made available weeks after the outbreak onset.[u]

From an economic point of view the two regions we have analyzed are similar.[v] Yet, they present notable differences as regards population density, relational density[w] and above all, the management of the health care system.

With Regional Law no. 19/2016, Veneto set up a new regional instrumental and centralized body named "Azienda Zero," which took over the planning, control and management of regional functions. In addition, the previous 21 Local Social Health Units (ULSS) were merged[x] into 9. The decision to establish Azienda Zero was taken with the aim to promote

[s]More specifically, the Lombardy Region re-organized regional health system services by issuing resolution No. XI/2906 of March 8, 2020, subsequently supplemented by decree No. 3353 of March 15, 2020.

[t]Resolution no. XI/2906 of 8 March 2020 reads: "Faced with the need to swiftly free up Intensive and Sub-Intensive Care beds and ordinary hospitalization beds for acute care hospitals, it has become necessary to make available to the Regional System all beds reserved for 'Out-of-hospital care' (subacute, post-acute, specialist health rehabilitation, pneumology especially, intensive and extensive intermediate care, Nursing and Residential Care Facility beds)."

[u]Reference is made to hospital facilities purposely built to cope with the Covid-19 epidemic, such as Bergamo Fair hospital or the Milan Fair hospital in Rho (https://www.ospedalefieramilano.it/it/index.html).

[v]The 2020 ISTAT annual report indicates that Lombardy has by far the largest number of residents in Italy, exceeding 10 million inhabitants (10,060,574). The region of Veneto, by comparison, has a number of residents equal to 4,905,854 people. Lombardy also has a higher population density than Veneto (420 inhabitants per square km, against 270). Both regions host international airports, are heavily involved in international trade and are tourist destinations. As such, they present similar risk factors for exposure to imported pathogens. The 11 well-being indicators set out by the OECD (Organization for Economic Co-operation and Development), including health, are virtually identical in the two regions, as are average age (45.9 vs 45.4 years) and life expectancy (84 years for both).

[w]For an in-depth discussion of mobility and commuting factors in the spread of Covid-19 see Chapter 3 of this volume.

[x]ULSS are the Local Social Health Units of the Veneto Region. They are bodies of the Italian public administration entrusted to the provision of health services.

savings and boost efficiency, leaving local health units (ULSSs) free to provide quality health care and services to citizens.[y] The aforementioned law provided for a 15% increase in the number of beds in community hospitals and a 60% increase in general practitioners. Such structural arrangement positively impacted regional performance, so much that Veneto was at the top of the LEA Grid, which classifies Italian regions based on their ability to guarantee essential care levels. It was probably this governance model that, at a time of emergency, placed less strain on the hospital system, favoring first-hand care via what is named "proximity medicine," that is the availability, throughout the region, of facilities and care professions able to support the medical, social, and care needs of citizens, with special attention to those who are not self-sufficient. Veneto's governance model seems to have responded well even during the Covid-19 epidemic.

In Lombardy, Regional Law no. 23/2015 established an organizational system based on three levels. The Region itself[z] maintained central planning, direction and control functions. The 15 ASLs (Local Health Authorities) were merged into eight new ATS entrusted with health planning, commissioning and monitoring functions. At the local level, the 27 Territorial Social Healthcare Companies (ASST) and the four Scientific Hospitalization and Care Institutes,[aa] were exclusively entrusted with operation management and therefore also the provision of hospital and territorial services. By simultaneously keeping alive both health agency models (ATS and ASST), Lombardy strove to approach a pattern of "integrated" governance.[ab] However, such integration has yet to be fully achieved. In fact, there is an at least partial overlap between the functions performed by the three levels of the regional health system; in particular, it is still partly undetermined whether agencies and ATS share some functions between planning and strategy. Furthermore, it is not yet clear whether ATS bear the same coordination responsibilities as ASSTs or whether they cover instead the role of arbitrators for a competitive contest between ASSTs. Ultimately, the regional health system of Lombardy seems indeterminate (Cantarelli et al., 2017, p. 374). Its setup has converged key activities such as health screening and acute pathology care onto large hospital complexes, relieving territorial facilities of these tasks and thus favoring an outpatient approach.

[y] For more information on the territorial setup of the Veneto system, you may consult the Region's website at: https://www.regione.veneto.it/web/guest/aziende-ulss-e-ospedaliere.

[z] The regional layout consists of the Unified Welfare Department (Assessorato Unico al Welfare) and its numerous agencies, such as the Control Agency, the Epidemiological Observatory, the Agency for the Promotion of the Lombardy Health Care System in the World, ARCA (Regional Central Purchasing Company), LISPA (Lombardia IT Services SpA), EUPOLIS (training and research agency), Finlombarda, Lombardia Infrastructure and Health Research Foundation (Cantarelli et al., 2017, p. 368).

[aa] Excellence hospitals which pursue research, mainly clinical and translational, in the biomedical field as well as in the field of the organization and management of health services and carry out specialty care and treatment or conduct other excellence-related activities. For more information, you may consult the Ministry of Health's webpage at: http://www.salute.gov.it/portale/temi/p2_6.jsp?id=794&area=Ricerca%20sanitaria&menu=ssn.

[ab] The term "integrated" governance refers to the fact that a single agency is both a lender and a producer of services. As a matter of fact, ATSs (Health Protection Agencies) in Lombardy do not provide health services directly, but manage contractual terms with health service providers (be they ASSTs or private agencies). De facto, this places public and private health care on the same level (Cantarelli et al., 2017, p. 369).

In the Covid-19 epidemic phase, this meant that local facilities throughout the region could not possibly contribute to an effective management of contagion. All infection cases converged instead onto hospitals, which were rapidly overburdened and thus also obliged for months to forgo scheduled interventions for other pathologies (Gardi et al., 2020, pp. 61–64).

The two regions are very similar in terms of the number of hospital beds for acute cases per 1000 inhabitants (3.05 in Lombardy versus 3.01 in Veneto). Per capita health expenditure is comparable, while the number of adults per GP is slightly higher in Lombardy (1400) than in Veneto (1342). In the public health sector,[ac] differences between the two regions are much greater: Lombardy has three public health laboratories (about 1 for every 3 million inhabitants), while Veneto has 10 (about 1 for every 500,000 inhabitants). Lombardy has eight public health prevention departments (1 for every 1.2 million inhabitants) against nine in Veneto (1 every 500,000 inhabitants). This unbalanced ratio between regional facilities and citizens results from the policies adopted by the two regions. Thus, in Lombardy, territorial medicine fails to play a mediation role between patient and hospital. Home care, which is meant to provide home services to the elderly, people with disabilities and people with chronic diseases, was extended to 3.5 people in Veneto (as of 2017) per 100,000 inhabitants and only half of that in Lombardy (1.4/100,000).[ad]

With regard to swab-testing in the two regions, the graphs in Fig. 5.9 and in Fig. 5.10 record data provided by the Ministry of Health regarding positive and negative swab tests in Lombardy and Veneto, thanks to which it was possible to chart daily data starting from February 24 until May 4, 2020.

Key provisions adopted in order to stem the spread of Covid-19 among the population are represented here along swab-testing. The policy adopted by the Veneto Region went immediately in the direction of monitoring the entire population and not only symptomatic cases, as shown by graph data. The outlook in Lombardy, on the other hand, featured marked variability, a slight upsurge trend and weekly cycles. Positive cases initially exceeded negative swab tests, which implies a chronic lack of total swab tests in virus monitoring. The graph shows that, at least initially, a large number of infected people were swab-tested in Lombardy.[ae]

[ac]By "public health laboratory" (LSP) we mean a facility that operates in the following fields: (i) chemical research in food, drinking water, or drugs; (ii) preventive microbiology; (iii) microbiology of food and water; (iv) biochemical human-matrix tests for preventive purposes and oncological screening; (v) epidemiological and biosafety investigations. The LSP is a specialized multifunctional facility which complements the tasks of ATS-provided services, with particular reference to the Services of the Department of Hygiene and Health Prevention, and covers the skills needed to carry out prevention-related tasks which involve laboratory interventions of considerable technical and professional complexity.

[ad]Data are provided by the latest "Monitoring of LEAs via the LEA Grid" of 2017, available at the link: http://www.salute.gov.it/imgs/C_17_pubblicazioni_2832_allegato.pdf.

[ae]Since, however, in nearly all cases swab-testing was carried out repeatedly on the same individuals, such data was included here solely as a distributive trend.

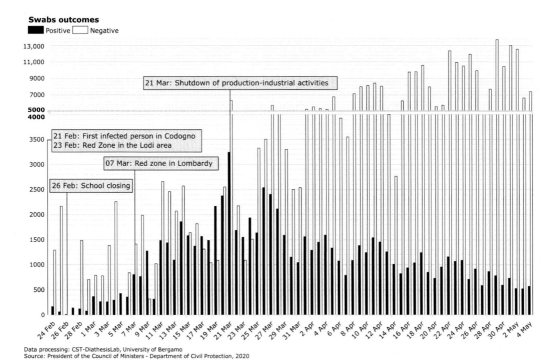

FIG. 5.9 Swabs carried out in Lombardy from February 24 to May 4, 2020.

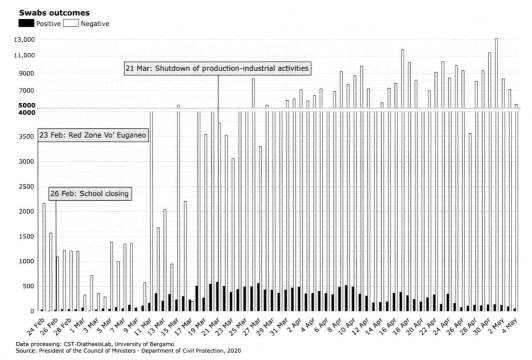

FIG. 5.10 Swabs made in Veneto from February 24 to May 4, 2020.

5.3.3 Conclusions

After an overview of fragmentary healthcare competences between State and Regions in Italy, our study has exemplified the consequences of such fragmentation in two Italian regions. Regional health systems do differ at several levels. They rely on different organizational setups and have dissimilar territorial impacts. The present study has addressed such issues with regard to the management of the Covid-19 epidemic. The Higher Health Institute[af] argues claims that the organization's role was crucial in monitoring new infections and breaking the Covid-19 chain of transmission. Based on their solid and effective presence on the territory, regional services were able to ensure key procedures such as contact tracing, management of reports, and home-administered swab testing assistance. The effectiveness of Covid-19 pandemic control measures cannot disregard the resources of tracking systems for promptly identifying and isolating contagious people and their contacts (Di Bari et al., 2020, p. 2). The organization of each regional health system and the policies that have been adopted since the onset for Covid-19 monitoring are therefore fundamental. Ultimately, excessive fragmentation of the National Health System has not eased the adoption of shared policies, which in turn affected data collection, prior to the planning of Covid-19 containment policies.

The Covid-19 emergency has exposed the shortcomings of a model based on a rigid separation of competences between the various levels of government involved in the management of a health crisis. A general lack of consistency in the applicability and application of national provisions eventually emerged. In the case of highly infectious contagion, the differences between the Italian regions have turned out to have a negative impact, because they hampered coordinated action for containing the Covid-19 epidemic.

[af]These statements may be found on the web page of the Higher Health Institute (Istituto Superiore di Sanità): https://www.epicentro.iss.it/en/coronavirus/sars-cov-2-ipc.

References

Binkin, N., Michieletto, F., Salmaso, S., Russo, F., 2020. Protecting our Health Care Workers while Protecting our Communities during the Covid-19 Pandemic: A Comparison of Approaches and Early Outcomes in Two Italian Regions. medRxiv. Preprint https://doi.org/10.1101/2020.04.10.20060707.

Cantarelli, P., Lega, F., Longo, F., 2017. La regione capogruppo sanitaria: assetti istituzionali e modelli organizzativi emergenti. In: Bocconi, C.E.R.G.A.S.-S.D.A. (Ed.), Rapporto Osservatorio sulle Aziende e sul Sistema sanitario Italiano. Egea, Milano, pp. 363–381.

Di Bari, M., Balzi, D., Carreras, G., Onder, G., 2020. Extensive testing may reduce Covid-19 mortality: A lesson from northern Italy. Front. Med. 7 (402), 1–5. https://doi.org/10.3389/fmed.2020.00402.

Franco, F., Di Napoli, A., 2019. Metodi di campionamento negli studi epidemiologici. G. Tec. Nefrol. Dial. 31 (3), 171–174.

Gardi, L., et al., 2020. La riforma del Sistema Sociosanitario Lombardo (LR 23/2015): analisi del modello e risultati raggiunti a cinque anni dall'avvio. AGENAS, Rome.

Ministero della Salute, et al., 2020. Test di laboratorio per SARS-CoV-2 e loro uso in sanità pubblica: nota tecnica ad interim. Istituto Superiore di Sanità, Rome.

Zucchi, A., Ciampichini, R., 2020. Analisi longitudinale per età dei positivi a tampone Covid-19 nella provincia di Bergamo. ATS Bergamo-Servizio Epidemiologico Aziendale.

Further reading

Anelli, S., et al., 2020. Emergenza Covid-19: studio del sistema dei ricoveri e delle risposte nei modelli organizzativi nelle diverse regioni italiane. Smart eLab 15, 1–16. https://doi.org/10.30441/smart-elab.v15i.170.

Ferrari, D., Cabitza, F., Carobene, A., Locatelli, M., 2020. Routine blood tests as an active surveillance to monitor Covid-19 prevalence. A retrospective study. Acta Biomed. 91 (3), 1–9. https://doi.org/10.23750/abm.v91i3.10218.

Scaccia, G., D'Orazi, C., 2020. La concorrenza fra Stato e autonomie territoriali nella gestione della crisi sanitaria fra unitarietà e differenziazione. In: Forum di Quaderni Costituzionali. Vol. 3, pp. 108–120.

Villa, M., 2020. Coronavirus: la letalità in Italia, tra apparenza e realtà. Italian Istitute for International Political Studies, Milano.

5.4

Dysfunctions and inadequacies in Health Districts and Nursing and Residential Care Facilities for the elderly in Lombardy as highlighted by the Covid-19 epidemic

Emanuele Garda

5.4.1 A model for elderly residential hospitality to the test

For some time, throughout the Italian territory and in individual regions[a] Nursing and Residential Care Facilities have been technically interpreted as non-hospital facilities intended to accommodate elderly non-self-sufficient persons[b] who cannot be directly assisted in their homes.[c] The purpose of RSAs, therefore, concerned the need to provide reception, health care, assistance, and recovery services in cases where it was recognized that it was

---

[a]Italy's administrative setup geographically comprises 20 Regions, which the Constitution defines as territorial entities endowed with their own statutes, powers, and functions. Different competences between the State and the Regions are in place with respect to handling of various issues. The "right to healthcare" falls within concurrent legislation and not in the matters of exclusive State competence.

[b]Berloto and Perobelli (2019) point out that the phrasing "non-self-sufficient elderly" includes people aged 65 or over, with functional limitations laid out in the definition of disability provided by the *International classification of functioning, disability and health* (ICF). These limitations may refer to: bed, chair, or house constraints; limitations in life functions and daily activities; issues with walking, using stairs and picking up objects from the ground; communication difficulties.

[c]Rotolo (2014) explains that, whenever home care is not possible or deemed inappropriate, non-self-sufficient persons are offered a number of dedicated residential or semi-residential care services. The former facilities (such as RSAs) have hotel-like features and host non-self-sufficient individuals either temporarily or permanently. The latter (semi-residential services, such as Day Care Centers) are services provided in day care facilities that support partially self-sufficient elderly people.

169

impossible to provide both continuous health treatments and specialist medical assistance directly at home.[d] This social welfare model was also the result of some social changes that took place in Italy. As the Italian Office for National Statistics pointed out recently, demographic, and social changes observed Italy in recent decades have altered family structures and the existing informal care network, usually consisting of family members, relatives, friends, or neighbors, which in the past allowed elderly people to live independently in their own homes (ISTAT, 2020, p. 126). Especially over the last decade, this new scenario and the growing "weaknesses" of the elderly population, to be ascribed not solely to age, have swayed welfare policies towards the introduction and consolidation of a range of care models. In the context of the Covid-19 epidemic, these may in fact have contributed to a swift viral spread, which exposed fragilities underlying our current patterns of living.[e]

Such varied context nonetheless confirmed that hospitalization in long-term nursing and residential care facilities of an elderly person or a person with severe physical or psychological disabilities ought to be endorsed only in the most complex cases, i.e., whenever the care load proves too high for the resources or technical-psychological support of a family network and the local health services. This was an essential pre-requisite in procedures for accessing these facilities, given the complexity of cases, the targeted assistance plans to be implemented for each person and, finally, the involvement of the National Health Service.

Since their inception, RSAs were set up as "specialized clusters"[f] within a range of organization and management models for patients, staff, and internal facilities, which were deployed across Regions,[g] in urbanized areas, or in settlement contexts tightly connected with urban centers, in accordance with national programs and regulatory guidelines. Such prerequisites, at least on a regulatory level, were laid out with the aim to contain all possible forms of isolation of the elderly and the potential difficulties with respect to maintaining the elderly's ties with their families.

Since RSAs host people who are physically or psychologically challenged with a view to managing chronic ailments for non-self-sufficient patients (Fara and D'Alessandro, 2015), the recent Covid-19 emergency in Italy and notably in Lombardy[h] has tended to turn such facilities into hotbeds of infection and elements of territorial weakness. The weakness of the

---

[d] These provisions refer to DPCM Decree dated 22 December 1989, available at the following link in the Italian Official Gazette: https://www.gazzettaufficiale.it/eli/id/1990/01/03/089A5965/sg.

[e] This section adopts the theoretical framework and the methodological approach laid out by Emanuela Casti in the introduction to this volume, which identifies health facilities, in particular those intended for the care of the elderly such as RSAs, as one of major weaknesses of current living patterns. In fact, RSAs seem to have played a key role during the "epidemic" phase, that is the phase of maximum viral spread in the spring of 2020, favoring a rapid and pervasive spread of the infection. For an in-depth treatment of this issue, see also: Casti (2020), especially page 66.

[f] RSAs provide patients with services that home care would find hard to implement, i.e., (i) medical assistance offered by doctors specialized in geriatrics and general medicine; (ii) continuous nursing care; (iii) rehabilitation assistance provided by professionals (physiotherapists, psychologists, educators); (iv) support and help guaranteed by social and health professionals.

[g] For an in-depth analysis of RSAs in Italian regions, see Chapter 5.2 in this volume.

[h] According to the Higher Health Institute (Istituto Superiore di Sanità, ISS), within the Italian territory, the highest percentage of deaths out of the total was recorded in Lombardy (41.4%), Piedmont (18.1%) and Veneto (12.4%) (ISS, 2020, p. 8).

people hosted in RSAs was, of course, aggravated by their age. According to the Italian Office for National Statistics (ISTAT, 2020, p. 126), the 247,000 elderly people hosted at the end of December 2018 in these facilities shared a clear demographic profile. On the basis of gender, the elderly population residing in Italy has a larger share of women (74%), with a very high average age (77% were over-80 years of age). Of these, over half were over 85 years old. The condition of "weakness," however, may not be ascribed solely to the age of the RSA subjects, even if age is one distinctive and significant factor. When applied to people hosted in RSAs, age must rather be taken as a partial indicator (Cesari and Proietti, 2020; Medford and Trias-Llimós, 2020) of the greater incidence of the epidemic and, above all, read within a broader and multi-factor framework. The deaths recorded in Italy among older patients fall within a context of both "multiple comorbidity" (Landi et al., 2020; Onder et al., 2020), and of the persistence, in affected sites, of other features that turned out to be problematic even in the initial phase of contagion. Such features may be considered as variables independent of the condition and will of an individual "inhabitant," because due to other, exogenous factors.

The surge in Covid-19 related deaths recorded in Italian RSA facilities over the past months, constantly monitored by the ISS,[i] has raised a number of critical concerns[j] which have been the subject of both heated political and institutional debate and of careful assessment on the part of judicial bodies. It is a set of concerns to do with intricate issues and possible liabilities in RSA management policies and the handling of Covid-19 patients during the most acute phase of the crisis. Thus, for instance, the decision to transfer patients affected by Covid-19, or patients with possible Covid-19 symptoms, to RSAs, whether to break the quarantine cycle or to conclude their convalescence, was aimed primarily at alleviating strain on hospital facilities. In retrospect, this decision made by some Italian regions was the equivalent of an "original sin," as Berloto et al. (2020) polemically put it: when the novel Covid-19 epidemic was found to pose a serious threat to the health of citizens, public bodies turned most of their attention to hospitals, and neglected nursing facilities, despite the obvious risks. Nor was this the only questionable decision[k]: a second policy seems to have affected the spread of contagion. As "semi-enclosed communities," RSAs usually admit controlled forms of interaction with the outside: the subjects they host are regularly in touch with their relatives, as well as with the specialized staff or medical staff employed in these facilities. The continued mobility of the latter groups (relatives and staff) between the inside and the outside may be seen as a tool "for maximizing social interaction" (Casti, 2020, p. 64). In the context of an ever-evolving epidemic on the outside, the unstable homeostasis of the RSAs' inner environment turns into an element of territorial "fragility." During the initial "onset phase," when little was known of

[i]Starting from 24 March 2020, the Higher Health Institute (Istituto Superiore di Sanità), has launched a survey on Covid-19 infection in RSAs in order to monitor the situation and adopt viable strategies. The survey covered 3292 RSAs (96% of the total) distributed throughout the national territory. Further information may be found on: https://www.epicentro.iss.it/en/coronavirus/sars-cov-2-survey-rsa.

[j]Errors in the initial management of the epidemic have been replicated by different countries: for instance, shortages in PPE supplies or delays in testing employees and patients (Inzitari et al., 2020).

[k]In the Lombardy Region this decision was made with a resolution of the Regional Council of 8 March 2020. Action taken by the Italian Regions in the initial phase of the epidemic was often fragmented, due to dissimilar RSA management policies for each regional territory. And fragmentation was exacerbated by the systemic lack of a national-level network of coordination (Berloto et al., 2020).

the new virus, the reticularity of RSAs, combined with an underestimation of actual risk and the absence of specific operational protocols, certainly contributed to the spread of Covid-19 inside these facilities. And in the subsequent phase, the "epidemic" phase, the same reticularity led to a dramatic surge of outbreaks and a rapid, severe spread of contagion (Casti, 2020, p. 65). The high mortality rates observed in some RSAs attest to a swift process of viral proliferation and concentration, whereby a combination of social and spatial features de facto turned these facilities into high-diffusion clusters of infection. To be sure, the role of RSAs in determining severe Covid-19 contagion was not limited to Italy. A rapid survey of international literature on the subject (Comas-Herrera et al., 2020; Inzitari et al., 2020) confirms that Nursing and Residential Care Facilities faced significant increases in deaths attributable to the spread of Covid-19.[1] In fact, at an international level, the spread of diseases in Nursing and Residential Care Facilities (Inzitari et al., 2020; Spasova et al., 2018) and, above all, the "fragility" condition we mentioned above, have been traced back to a core set of reasons (Chin-Cheng et al., 2020), namely the facts that: (i) RSA residents have multiple chronic diseases; (ii) residents share the same "resources" and the same medical care (provided by medics or paramedics) which may introduce and consequently facilitate the transmission of infections; (iii) visitors, workers[m] and, in some cases (i.e. for partially self-sufficient elderly people), residents may enter and leave the facilities without limitations, becoming vehicles for the introduction of pathogens into these environments.

5.4.2 The national and Lombard care-system for non-self-sufficient people

Nursing and Residential Care Facilities have spread throughout the Italian territory since the nineties, following a set of regulatory initiatives at the national level such as Law no. 67 of 1988 and the Decree of the President of the Council of Ministers (DPCM) of 22 December 1989. The former measure was especially influential, since it launched and funded a multi-year program of interventions aimed at the construction of residences for the elderly and non-self-sufficient people. The DPCM, on the other hand, provided the Regions with guidelines for coordinating administrative activities related to the design and construction of RSAs. This initial set of guidelines was one indispensable reference tool for framing specific technical-descriptive features to do both with the dimensional and functional organization of the facilities (e.g., type and size of spaces, creation of residential or social activity areas), and with their ordinary management (e.g., usability of spaces, safety, and similar).

Despite this shared set of common measures, the framework of the Italian health system, including home care and, above all, facilities for long-term hospitalization or the hospitality of elderly persons, is currently marked by strong regional diversification (Volpato et al., 2020). Fragmentation and imbalance between regions are evident, even though it can be traced back

---

[1]Besides data culled from the local and national press of many countries, first-hand scientific evidence of Covid-19 spread in residential care facilities at an international level was put forward by the *International Long-Term Care Policy Network* and published in Comas-Herrera et al. (2020).

[m]New scientific evidence has confirmed that RSA staff may already be infected with the Covid-19 virus and hence be contagious, even when ostensibly asymptomatic (D'Adamo et al., 2020).

to a shared model in the care of non-self-sufficient elderly persons. In the regional supply chains, the set of care services provided is ideally tied to a network available to families and the elderly, whose individual components may interact in different modes based on changing needs over time (Berloto and Perobelli, 2019). Because of this, Italian Regions, including the specific case discussed in the following pages, operate within a fairly heterogeneous context in terms of a geography of supply and demand for needs-appropriate services, whereby the network of residential and semi-residential facilities is taken as a step following which follows home care,[n] but precedes residential hospitalization.

In the case of the Lombardy Region[o] two key features may be underscored with regard to the role RSAs played[p] in determining the intensity and severity of the epidemic. RSA presence within Lombardy is undoubtedly high[q]: 717 RSAs currently cover over 64,000 bed units, available either in public facilities (54 facilities), or in accredited private facilities (663 facilities) (Fig. 5.11). When cross-referenced to the elderly population present in the health districts established by regional legislation, this high availability of beds revealed marked disparities between the regions. In districts clustered around the central section of Lombardy, those linked to the territories of the Metropolitan City of Milan or to the provinces of Monza and Brianza, Bergamo, and Brescia (excluding the Brescia valleys), the ratio between RSA beds and the elderly population amounts to 2.5.[r] For the territories located in the northern sections (part of the provinces of Brescia, Sondrio, Lecco, and Como) and southern sections (part of the provinces of Pavia, Lodi, and Cremona), this value shows a slight increase, exceeding the value of 5 beds per 100 elderly inhabitants. This data suggests that in the wide

[n] According to Gori and Rusmini (2015), the stability of the Italian health system in terms of support for the elderly is based on two different approaches: the first refers to the self-organization of families who turn either to personal assistance services or to other more or less formalized home care solutions; the second approach is that of resorting to health services (for patients with acute pathologies) in order to contain emergency situations or situations of temporary health care which limit the self-management of families and involve repeated hospitalizations or prolonged hospital stays.

[o] Located in the north of Italy, between the mountain range of the Alps and the Po river, the Lombardy Region covers an important geographical and administrative area. With an area of about 24,000 km^2 and nearly 10 million inhabitants, it is the third most populated region in Europe after Île-de-France and Baden-Württemberg, with approximately 2.3 million residents over the age of 65.

[p] RSAs in Lombardy have developed since the nineties, following Decree 1 in 1986, the Social Health and Care Regional Plan (1988/1990 3-year period) and the Elderly Care Project (1995–1997 over a 3-year period). Thanks to these initiatives and to the consolidation of contracting criteria for the provision of services by private individuals, the Lombardy model was strengthened over time and currently records figures that have no equal on the national scene. According to the Higher Health Institute (Istituto Superiore di Sanità, 2020), Lombardy had a number of RSA residents equal to 26,981 as of February 1, 2020. Veneto trailed significantly behind that, with 17,381 residents, followed by Piedmont with 16,629 residents, even though these are both populous regions, with over 4 million total residents.

[q] These are the Nursing and Residential Care Facilities accredited to regional bodies, according to current legislation, and recorded in official data posted in the Lombardy Region Open Data database (www.dati.lombardia.it).

[r] This value refers to the number of beds available in RSAs for every 100 inhabitants over 65 years of age.

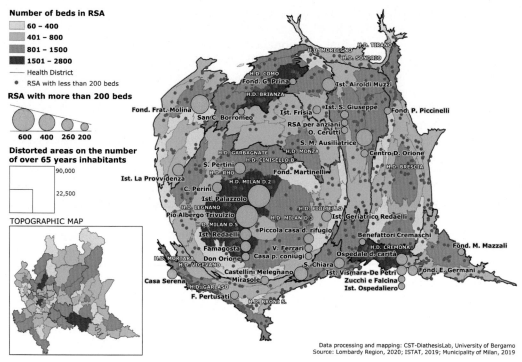

FIG. 5.11 Distribution of RSA in Lombardy, scaled to the number of beds.

urban regions along the central strip of Lombardy, featuring generally high population density[s] and long-term densification processes (Lanzani, 1997), RSA demand is limited. It also confirms that in densely populated districts, residential care demand far exceeds availability, so much so that families have repeatedly been found to look for alternative elderly care arrangements beyond their regional area of residence.

Secondly, from a regulatory point of view these facilities fall under complex set of regulations, which has undergone profound changes over the years. Following a lengthy process of regulatory reform (Regional Law No. 23 of 2015), this framework changed in 2015, which saw an overhaul of the healthcare system in Lombardy. The key principles of this legislative review, which confirmed a few cornerstones of previous laws, concerned: (i) the patient's freedom of choice regarding access to treatment; (ii) competitiveness between public and private facilities (based on an accreditation system); (iii) strengthening of an accountability separation between planning and services provision. The accreditation system and the consequent involvement of private facilities in the provision of health and of social or social-care services either for the non-self-sufficient elderly or for other service types is a fundamental component for guaranteeing the operation of the regional supply chain dedicated to the care and hospitality of these people. The data presented at the beginning of this section are telling indicators of this "imbalance" in favor of private operators and of the essential role they have achieved.

[s]For an overview of densities across Lombardy, see Chapter 1.3 of this volume.

Another change came with the establishment of Health Protection Agencies or public facilities responsible for implementing the plans and policies of the Lombardy Region, via the provision of health and social and health services through accredited bodies and contractors (both public and private) which impacted the patients' freedom of choice (with respect to access to treatment and the increase in the existing offer). With the reform, the territorial catchment of each Agency was divided into Districts based on a demographic criterion, that is, based on a population count exceeding 80,000 inhabitants. For high population density areas, this criterion was raised to 120,000 inhabitants, while in mountain regions, or regions with low population density, it was reduced to a minimum population count of 25,000 inhabitants. Both public and private RSAs in Lombardy deal directly with the framework described above, providing their specific services on the basis of both an indispensable regional accreditation process and the signing of a contract with the respective ATSs to which they belong. These conditions are necessary in order to guarantee partial coverage of hospitality costs by the Regional Health Fund, with the remaining amount charged to the guests themselves or to the Municipalities of residence.

5.4.3 Lombardy in the plural

As mentioned above, the high-density network of RSAs in Lombardy comprises >700 facilities which include both public (54) and private (663) institutions. The strong asymmetry between these two groupings may be ascribed to the accreditation mechanism introduced by the Lombardy Region Council, who adopted a principle widely used in many components of the healthcare system, with the aim of involving all those agencies which had long been active in the hospitality sector for the elderly at the local level. In addition to fulfilling a demand for healthcare and bed units which would otherwise have required significant public investment, this choice empowered a vast and heterogeneous constellation of facilities which could contribute specific knowledge and experience in the care of the elderly and had strong local roots.

The heterogeneity of RSAs is not limited to their number, but concerns other aspects as well. The legal status of RSA managers confirms the varied pattern of the Lombard RSA system. Monitoring put in place by the Lombardy Region indicates that the most widespread type of agency is the Foundation, to which 306 facilities may be ascribed. Other agency types trail well behind, for example, limited liability companies (90 facilities), cooperation and social solidarity societies (81 facilities), ecclesiastical bodies (26 facilities) and non-profit organizations under private law (26 facilities).[t] In simplified terms, this framework may be said to consist of dichotomies: the one already mentioned between public and private RSAs and a possible division between active subjects with "not-for-profit" purposes which apply to 55% of the total on the one hand, and "for profit" agencies more clearly oriented towards business on the other.

The many Nursing and Residential Care Facilities are spread out throughout Lombardy in accordance with the propensities and "styles" which refute an ideal modeling in terms of isotropic space, hard to find within a heterogeneous and plural territory (Lanzani, 1991),

---

[t]Classification proposed by the Lombardy Region presents 27 different categories which cover a wide range of sizes, ownership types or purposes among subjects involved in the RSA system.

and allow instead for a number of alternative models. The outline of such model in the pages that follow rests on an awareness of the limits and risks involved in pure geolocalization analysis, which calls instead for comparison with other "dimensions" in an attempt to recover the "social sense of the territory" (Casti, 2015).

Fig. 5.11 shows the distribution of Nursing and Residential Care Facilities in Lombardy within separate health districts. Specifically, map ground anamorphosis outlines the boundaries of health districts (thin black line) and of provinces (thick black line) distorted to reflect the number of elderly people (residents over 65 years of age) in each of them. Anamorphosis makes it possible to record spatial information as a function of social data, in this case data related to the elderly population. The area around Milan has the highest concentration of elderly residents, whose presence is instead considerably less in mountain areas such as the province of Sondrio and the lower Lombard plains as in the provinces of Pavia and Lodi. Social data are also color-coded in shades of green to reflect the number of bed units in each RSA, with a high concentration both in the north-west area of Milan (Milan Health Districts n. 2 and 5) and in other areas such as Cremona (Cremona Health District) and east Como (Brianza Health District). In order to highlight the contrast between large-sized care facilities (in term of bed units) and other medium-to-small-size facilities, two different visualization models were adopted. Specifically, RSAs with <200 bed units were marked via a uniform graphic symbol (orange dot) and on the basis of geolocation; facilities with more than 200 bed units were linked to their respective districts via a graphic symbol that is not geolocated and varies in size according to the number of beds (scaled aerogram in yellow).

Overall, the adoption of a reflexive approach in mapping makes it possible: (1) to highlight the social features of territory via anamorphic distortion of Lombardy's Health Districts; and (2) to provide integrated datasets for an articulate interpretation of phenomena. At the same time, reflexive cartography gives us an immediate grasp of the widespread distribution of RSAs in Lombardy and of their problematic presence as fragilities of contemporary living within urban contexts. In particular, comparison between the distribution of RSAs,[u] the geography of different health districts outlined by the latest healthcare reform and population data for over 65 s (in relation to District boundaries) provides evidence of key features for Lombard context. The overall chart in Fig. 5.11 underlines the weight of the Municipality of Milan,[v] whose Districts two, three and five host several large-size facilities. These are hospitalization facilities that host well over 200 beds. In at least two cases, which also featured

---

[u]Information on accredited Nursing and Residential Care Facilities was obtained by re-processing data available in the Open Data Database of the Lombardy Region (www.dati.lombardia.it).

[v]Because of its high number of residents (a count of 1,404,431 inhabitants as of 31/12/2019 according to data from the Municipal Registry Office), the territory of Milan was broken down into five Health Districts. This data ranks Milan first in Lombardy, but only second in the overall Italian ranking of large cities, after Rome (2,837,332 according to ISTAT in 2019) and before Naples (962,589 according to ISTAT in 2019). Nonetheless, crucial differences remain between the municipalities of Rome and Milan with respect to administrative boundaries and population density. Rome covers an area of 245 sq. km with a population density of nearly 2200 (inhabitants/sq. km); Milan covers an area of 181 sq. km with a population density of nearly 7700 (inhabitants/sq. km). Albeit necessarily synthetic, such data confirm that Milan is a city constrained within a restricted administrative geography (Bolocan Goldstein, 2009), the bearer of an exceptionally dense urban environment.

leading Institutes for elderly care and hospitalization, hospitality capacity reached nearly 600 beds. State-of-the-art institutes include the Palazzolo Institute, located in District two, and the Pio Albergo Trivulzio Institute, located in District five, which also hosts other major facilities such as the Redaelli, Famagosta and Don Orione institutes. Facilities like these are not important solely by virtue of the number of bed units for non-self-sufficient people they provide. Rather, they matter because they are strongly rooted in the local territory[w] and they are multifunctional facilities ideally positioned to offer a range of services for the elderly.[x] The distribution of bed units within the Lombard healthcare districts further confirms the weight of the region's capital, Milan, especially of its western sector, which, in the map, ranks highest in terms of beds (between 1501 and 2800 units). Anamorphic distortion obtained by altering health district boundaries according to the number of elderly people (over 65) emphasizes the significance of this social component relative to the urban population and may lead to a keener understanding of possible causes for the quantity and concentration of bed units compared to the rest of Lombardy. The map also gives an immediate perception of the impact of an elderly population, relative to a local supply of bed units which, over the years, has had to adapt to the growing needs of an "aging society" (Fara and D'Alessandro, 2015).

In other Lombard districts, the presence and distribution of RSAs generates different localization geographies that may be traced to other spatial models. Single large-sized homes in terms of offer and reception capacity (which exceed 200 bed units) may therefore be found alongside a large number of medium and small size Nursing and Residential Care Facilities. This would include lower urbanization districts that do not match the figures for Milan or Brianza, but involve province capitals such as Bergamo, Lecco, Varese and, albeit in limited terms, Mantua. As in the case of Milan, anamorphic visualization for these large cities emphasizes polarities relative to population aging and the provision of an increasingly adequate offer within a regional framework. The model may include Districts belonging to the central conurbation of Milan, characterized by the presence of medium-sized administrative clusters such as Cinisello Balsamo (about 76,000 inhabitants), Pioltello (about 37,000 inhabitants), Garbagnate Milanese (about 27,000 inhabitants) and Rho (about 50,000 inhabitants). To recall a keen analysis by Glaeser (2011), such clusters may be said to reflect the profoundly social nature of cities, both in terms of population density and intensity in interpersonal and economic relationships. They may be described as spaces of reticular and polycentric urbanity (Casti, 2020). In addition, the infrastructural endowment and the availability of a complex private and public mobility system in these cities is a feature of some importance with respect to the location of RSAs, especially those of large size. This would confirm the principle of "easy accessibility" evoked by the initial regulatory measures discussed earlier.

Another model refers to district contexts characterized by a greater propensity for polycentrism but without strong asymmetries among RSAs. It is the case of facilities that host less

[w]The Pio Albergo Trivulzio, in operation for more than two centuries, endeavors to support the most disadvantaged elderly and has strong historical and political ties to the city of Milan.

[x]Having been set up as multi-layered organizations that provide a gamut of health services for self-sufficient and non-self-sufficient elderly people, the two facilities mentioned here, like many others in Lombardy, may be considered as "complex machines" (in terms of the size of their buildings, of spaces, functional diversification, and personnel management) and as prominent nodes of the urban fabric with a strong propensity for multiscalarity.

than 200 bed units but are located in prominent areas by virtue of the high number of elderly inhabitants or the dense urbanization level (over 40% of the municipal area). This trend may be recognized, for instance, in the healthcare districts of Brescia, Monza, Legnano, Vigevano, and Como, or in District number three of the Municipality of Milan.[y] Similar considerations apply to the Districts within the province boundaries of Monza and Brianza, which feature a settlement framework which bears the distinctive features of other highly urbanized areas of Lombardy. Since the 1990s, the large polycentric conurbation that runs within this territory and includes municipalities of sizable resident population,[z] has been conventionally identified as a highly urbanized and reticular "urban environment," comparable to the urban area of Milan in terms of land coverage and building density (Boeri et al., 1993). The lack of large facilities within these Districts, similar in terms of urbanization and importance to those mentioned for the previous model (and often tied to Province capital cities) often depends either on the architectural setup of RSAs in those areas, or on the policy of favoring sharper separation between facilities devoted to the non-self-sufficient elderly (RSAs proper) and other facilities for those who are self-sufficient.

The same polycentrism, albeit with slight variations, may be found also in other urban clusters, which are smaller in size because they catch less densely populated municipal administrations, such as in the districts of the northern sectors of Lombardy, or, on the contrary, in areas of the lower Lombard plain. This applies to some Districts within the boundaries of the province of Sondrio (with a count of nearly 180,000 inhabitants over 77 municipalities), for example Morbegno, Sondrio and Tirano, or to some Districts of the province of Pavia (with an overall count of about 547,000 inhabitants over 185 municipalities). Similar examples exist for the province of Pavia in the Districts of Broni-Stradella, Garlasco and Mortara. In this case, as evidenced by cartographic anamorphosis, low-density settlements (either demographic or relative to land coverage), morphological and orographic features of territories and the presence of a relatively smaller elderly population converge into a spatial model that attests to the spread of multiple smaller-size RSAs.

One final and unusual case is the one of Health Districts around the town of Cremona. From the point of view of settlements, such districts chart a conventional low-density development model (Lanzani, 1997) featuring point-like and multicentric urbanization which typically characterize territories affected by urban sprawl (Gibelli, 2002).[aa] This area hosts comparatively more agricultural space than other smaller urbanized clusters in Lombardy, but also features a high number of bed units in RSAs (the District of Cremona in fact ranks first on the map for number of bed units), as well as larger-size facilities. With respect to this

[y] In these areas, elderly population percentages out of the total resident population are as follows: 24% for Brescia, 24% for Monza, 20% for Legnano, 23% for Vigevano, 20% Como, and 22% for Milan (source: Regional Statistical Yearbook, 2020).

[z] High population density in provincial areas is a well-known feature of Italian demographics, with an overall Province count of 2166 inhabitants/km^2 second only to the Province of Naples with its 2615 inhabitants/km^2. Telling examples of this phenomenon are the municipalities of Monza (124,056 inhabitants), Lissone (46,445 inhab.), Seregno (45,447 inhab.), and Desio (41,997 inhab.).

[aa] According to Gibelli (2002, p. 17), urban sprawl refers to "an urbanization model which has low relative density, is stretched to the outer edges of the metropolitan region, has high land coverage, and outlines a discontinuous, if tendentially segregated territory optimized for single-purpose activities, mostly dependent on automobile transport."

second feature, cartographic data on the map underscore the Foundation Hospital of Charity (300 beds) in the Municipality of Casalbuttano (about 3800 inhabitants), the Vismara-De Petri Institute (267 units) in the Municipality of San Bassano (about 2220 inhabitants) and the Zucchi-Falcina Institute (213 beds) in the Municipality of Soresina (about 8900 inhabitants). One notable feature of these districts is that they host centuries-old foundations strongly tied to the local territory, often the results of donations from benefactors in the area and of the integration of charities, non-profit organizations, and charitable institutions.

5.4.4 Conclusions

This contribution addressed the broad topic of Nursing and Residential Care Facilities, most notably of facilities intended for non-self-sufficient subjects. A few key issues surrounding this complex area of research were discussed, especially touching upon recent calls for a holistic approach, with the aim to trace correlations in the ongoing debate over RSAs which followed the spread of the Covid-19 epidemic. We underlined how the notion of "fragility," typically ascribed to the category of "inhabitants" hosted in RSAs for the elderly, should not be traced exclusively to advanced age. Albeit certainly relevant, age-related considerations must be seen within a broader framework of endogenous and exogenous factors. Among endogenous factors, the coexistence of different pathologies, often in the chronic form, otherwise known as "multiple comorbidity" must be mentioned. For in the case of RSAs, comorbidity is in fact one of the prerequisites for access to such facilities. Among exogenous factors, we would include a set of issues that presumably played a central role in the spread of Covid-19 in RSAs, namely: (i) the need to share common spaces, therapies and, above all, medical and paramedical personnel. In an initial phase of limited knowledge about Covid-19 infection, all these acted as vehicles of viral transmission; (ii) interaction with the outside environment through frequent meetings with family members. While essential in psychological terms, these also facilitated contagion; (iii) lack of protocols or specific guidelines for the management of unexpected or exceptional events.

A second issue of national interest, but closely tied to lines of international research, concerns the need to rethink operational and organizational approaches applied inside facilities meant for elderly populations, and notably for non-self-sufficient people. As Barnett and Grabowski (2020) noted for the specific case of Seattle, the virulence of the Covid-19 epidemic may be taken as a sort of "Ground zero" for retirement homes. Also, such remarks must necessarily relate to the "global challenge" long imposed, especially on the West, by a protracted "phase"[ab] of progressive aging[ac] in societies and their populations (Partridge et al., 2018) and the consequent numerical weight non-self-sufficient elderly inhabitants have reached.

[ab] According to UN analyzes and projections, the average age of the world population was 26 years in 1990 but is estimated at 38 years of age by 2050, with a growth equal to 46% in 60 years (UN, 2013). The European Union has recognized that the percentage of over-65s, equal to 14.6% in 1990, and up to 17.2% in 2005, will be 19.4% in 2015 and 20.7% in 2020 (EC, 2014).

[ac] The phrase "societal aging" may be taken to cover all the social changes brought about on population because of the combined pressure of some phenomena (Fara and D'Alessandro, 2015), namely (i) reduction of birth rates; (ii) increase in infant mortality; (iii) increase in life expectancy at birth.

The aforementioned research should aim at introducing new protocols with immediate effect, due to the high risk of contagion affecting both the elderly hosted and the staff employed in RSAs (Gardner et al., 2020). Measures to combat Covid-19 should, for instance, focus on rapid testing for tracing or monitoring infections, as well as on the continuous training of medical and paramedical personnel (adequately safeguarded by PPE—Personal Protective Equipment). In addition to short-term interventions, essential to limit the spread of Covid-19, other much more complex issues should be addressed, starting with the need to promote greater coordination of Nursing and Residential Care Facilities at the territorial level. We should not forget that the lack of coordination, fragmentation of competences, and mismatch between policies (e.g., between social aspects and health measures) are in fact common traits to many European countries (Spasova et al., 2018). The Covid-19 epidemic has exposed the severity of all such shortcomings. As such, the epidemic marks an opportunity for reviewing existing models[ad] and for reassessing the colossal sprawl of some RSA facilities. The deaths internationally recorded in RSAs have rekindled an interest in the values of attentive care, quality workforce and high living conditions which ought to guide RSA policies (Inzitari et al., 2020).

Our analysis then outlined a typology of RSA residences for non-self-sufficient elderly persons in Lombardy and discussed their impact on the spread of Covid-19 contagion. As such, RSAs represent features of territorial weakness. In particular, through the use of reflexive cartography, we highlighted the close link that exists between some regional areas affected by high-density settlements and the regional system of RSAs. This applies, first of all, to the city of Milan where, in addition to a very large number of RSAs (60 in total), we recorded the presence of considerable size facilities in terms of the number of bed units available (two facilities host 600 bed-units). Anamorphic mapping provides a very effective graph of the two factors that characterize Milan: high population density and greater incidence of the elderly population. The same type of mapping highlighted the importance of other provincial capitals (e.g., Bergamo, Lecco, Varese, and Cremona), and of other medium-sized municipalities located within the Metropolitan City of Milan (e.g., Rho, Cinisello Balsamo and Pioltello), that is within a highly urbanized and reticularity-based model of areas around Milan. In the case of less-urbanized areas (e.g., in the districts of Pavia or in mountain contexts) where the elderly population is considerably lower than in other areas, a polycentric model may be said to emerge, due to the widespread distribution of smaller RSA facilities (under 200 bed units). The distributive geography of RSAs and the different models here discussed reconfirm and strengthen our hypothesis of a "plural" Lombardy, that is, a territory characterized by a great variety of scenarios and local situations, whose origin must be sought in the profound transformations that affected settlement and social policies after World War II (Tosi and De Carolis, 1990). Differences in the spatial distribution and size of RSAs respond both to the market-driven adaptation to a growing demand for elderly hospitality services in local contexts, and to the direct relationship with settlement environments, which present a variety of morphological and social networks (Palermo, 1997) across Lombardy. Such distributive setup, which relies on a system supported by both public and private facilities via regional

---

[ad] An interesting line of research is set out by Nacoti et al. (2020), according to which while "Western health systems have been built around the concept of patient-centered care," however, as authors state "an epidemic requires a change of perspective towards a concept of community-centered care."

accreditation programs, may not be seen as the result of holistic planning or as a manifestation of unified governance policies. What we are faced with is, rather, a heterogeneous, fragmented and potentially dynamic setup which, in Italy as in other international contexts, has been left to cope with an epidemic without adequate technical, cognitive, and operational tools.

References

Barnett, M.L., Grabowski, D.C., 2020. Nursing homes are ground zero for COVID-19 pandemic. JAMA Health For. 1 (3), e200369. accessed December 2020 from https://jamanetwork.com/channels/health-forum/fullarticle/2763666.

Berloto, S., Perobelli, E., 2019. Dalla Silver Economy al settore LTC in Italia: dati, sviluppi e tendenze in atto. In: Fosti, G., Notarnicola, E. (Eds.), Il futuro del settore LTC Prospettive dai servizi, dai gestori e dalle policy regionali. 2° Rapporto Osservatorio Long Term Care, EGEA, Milan, pp. 17–54.

Berloto, S., Notarnicola, E., Perobelli, E., Rotolo, A., 2020. Italy and the COVID-19 long-term care situation. In: Country report in LTCcovid.org, International Long Term Care Policy Network. CPEC-LSE.

Boeri, S., Lanzani, A., Marini, E., 1993. Il territorio che cambia. In: Ambienti, paesaggi e immagini della regione urbana milanese. Abitare Segesta Cataloghi, Milan.

Bolocan Goldstein, M., 2009. Geografie milanesi. Maggioli Editore, Santarcangelo di Romagna.

Casti, E., 2015. Reflexive Cartography. A New Perspective on Mapping. Elsevier, Amsterdam.

Casti, E., 2020. Geografia a 'vele spiegate': analisi territoriale e mapping riflessivo sul COVID-19 in Italia. Documenti Geografici 1, 61–83.

Cesari, M., Proietti, M., 2020. Geriatric medicine in Italy in the time of COVID-19. J. Nutr. Health Aging 24 (5), 459–460. https://doi.org/10.1007/s12603-020-1354-z.

Chin-Cheng, L., et al., 2020. COVID-19 in long-term care facilities: an upcoming threat that cannot be ignored. J. Microbiol. Immunol. Infect. 53, 444–446. https://doi.org/10.1016/j.jmii.2020.04.008.

Comas-Herrera, A., et al., 2020. Mortality associated with COVID-19 outbreaks in care homes: early international evidence. In: LTC Responses to Covid-19. International Long-term Care Policy Network. accessed December 2020 from: https://ltccovid.org/2020/04/12/mortality-associated-with-covid-19-outbreaks-in-care-homes-early-international-evidence/.

D'Adamo, H., Yoshikawa, T., Ouslander, J.G., 2020. Coronavirus disease 2019 in geriatrics and long-term care: The ABCDs of COVID-19. J. Am. Geriatr. Soc. 68, 912–917. https://doi.org/10.1111/jgs.16445.

European Commission, 2014. The 2015 ageing report—Underlying assumptions and projection methodologies. Joint Report prepared by the European Commission and the Economic Policy Committee. https://doi.org/10.2765/76255.

Fara, G.M., D'Alessandro, D., 2015. L'invecchiamento della popolazione: riflessi sulla soddisfazione delle esigenze socio-assistenziali. Techne. J. Technol. Architect. Environ. Architect. Health Educ. 9, 21–26.

Gardner, W., States, D., Bagley, N., 2020. The coronavirus and the risks to the elderly in long-term care. J. Aging Soc. Policy 32 (4–5), 310–315. https://doi.org/10.1080/08959420.2020.1750543.

Gibelli, M.C., 2002. La dispersione urbana in Europa. In: Camagni, R., Gibelli, M.C., Rigamonti, P. (Eds.), I costi collettivi della città dispersa. Alinea, Florence, pp. 15–24.

Glaeser, E., 2011. Triumph of the City. Penguin Press, London.

Gori, C., Rusmini, G. (Eds.), 2015. La rete dei servizi sotto pressione. In: Network Non Autosufficienza, V Rapporto sulla Non Autosufficienza. Maggioli Editore, Bologna.

Inzitari, M., Risco, E., Cesari, M., Buurman, B.M., Kuluski, K., Davey, V., Bennet, L., Varela, J., Prvu, B., 2020. Nursing homes and long term care after COVID-19: a new era? J. Nutr. Health Aging. https://doi.org/10.1007/s12603-020-1447-8.

ISTAT, 2020. Rapporto annuale 2020. La situazione del Paese. accessed December 2020 from https://www.istat.it/it/archivio/244848.

Istituto Superiore di Sanità, 2020. Survey nazionale sul contagio COVID-19 nelle strutture residenziali e sociosanitarie. Report finale. accessed December 2020 from https://www.epicentro.iss.it/coronavirus/sars-cov-2-survey-rsa.

Landi, F., et al., 2020. The new challenge of geriatrics: saving frail older people from the SARS-COV-2 pandemic infection. J. Nutr. Health Aging 24, 466–470. https://doi.org/10.1007/s12603-020-1356-x.

Lanzani, A., 1991. Il territorio al plurale. Interpretazioni geografiche e temi di progettazione territoriale in alcuni contesti locali. Franco Angeli, Milan.

Lanzani, A., 1997. Geografie degli ambienti insediativi lombardi. In: Palermo, P.C. (Ed.), Linee di assetto e scenari evolutivi della regione urbana milanese. Atlante delle trasformazioni insediative. Franco Angeli, Milan, pp. 27–38.

Medford, A., Trias-Llimós, S., 2020. Population age structure only partially explains the large number of COVID-19 deaths at the oldest ages. Demogr. Res. 43, 533–544. https://doi.org/10.4054/DemRes.2020.43.19.

Nacoti, M., Ciocca, A., Giupponi, A., Brambillasca, P., Lussana, F., Pisano, M., Goisis, G., Bonacina, D., Fazzi, F., Naspro, R., 2020. At the epicenter of the Covid-19 pandemic and humanitarian crises in Italy: changing perspectives on preparation and mitigation. In: NEJM Catalyst Innovations in Care Delivery. Massachusetts Medical Society, Waltham, pp. 1–5. accessed December 2020 from https://catalyst.nejm.org/doi/full/10.1056/CAT.20.0080.

Onder, G., Rezza, G., Brusaferro, S., 2020. Case-fatality rate and characteristics of patients dyingin relation to COVID-19 in Italy. JAMA 323 (18), 1775–1776. https://doi.org/10.1001/jama.2020.4683.

Palermo, P.C., 1997. Ambienti insediativi e processi di trasformazione. Verso nuove immagini e interpretazioni della regione urbana milanese. In: Palermo, P.C. (Ed.), Linee di assetto e scenari evolutivi della regione urbana milanese. Atlante delle trasformazioni insediative. Franco Angeli, Milan, pp. 7–26.

Partridge, L., Deelen, J., Slagboom, P.E., 2018. Facing up to the global challenges of ageing. Nature 561, 45–56. https://doi.org/10.1038/s41586-018-0457-8.

Rotolo, A., 2014. Italia. In: Fosti, G., Notarnicola, E. (Eds.), Il Welfare e la Long Term Care in Europa. Modelli istituzionali e percorsi degli utenti. Egea, Milan, pp. 93–115.

Spasova, S., Baeten, R., Coster, S., Ghailani, D., Peña-Casas, R., Vanhercke, B., 2018. Challenges in long-term care in Europe. In: A Study of National Policies. European Social Policy Network (ESPN), European Commission. Brussels.

Tosi, A., De Carolis, G., 1990. Lombardia. In: Astengo, G., Nucci, C. (Eds.), IT.URB.80 Rapporto sullo stato dell'urbanizzazione in Italia. Urbanistica informazioni, Rome, pp. 31–41.

United Nations, 2013. World Population Ageing 2013. Department of Economic and Social Affairs Population Division, https://doi.org/10.18356/e59eddca-en.

Volpato, S., Francesco, L., Antonelli Incalzi, R., 2020. A frail health care system for an old population: lesson form the COVID-19 outbreak in Italy. J. Gerontol. A Biol. Sci. Med. Sci. 75 (9), 1–2. https://doi.org/10.1093/gerona/glaa087.

Chapter 6. Public policies for epidemic containment

CHAPTER

6.1

Public policies for epidemic containment in Italy

Fulvio Adobati, Emanuele Comi, and Alessandra Ghisalberti*

6.1.1 Premise

In line with our research methodology, reconnaissance, and assessment of the range of measures adopted to counter Covid-19 outbreaks are conducted on a multiscalar architecture: from European scale to Italian national scale, to a breakdown analysis for the separate Italian region.

Specifically, guiding criteria for assessing the adoption of restrictive measures in the event of contagion or, alternatively, for relaxing measures in a European context may be said to fall within three categories (EU, 2020): (i) epidemiological criteria indicating the number of hospital admissions and/or new cases of contagion over an extended period of time; (ii) capacity and infrastructural endowment of healthcare systems; (iii) monitoring effectiveness, also in terms of large-scale diagnostic capabilities that make it possible to quickly identify and isolate infected people and in terms of track and trace strategies.

Such measures entail a system of rules, which over time has been enforced with varying degrees of severity, from informal recommendation to strict imposition, as regards device use and personal behavior. These guidelines are effectively condensed in the three catchwords promoted by the UK government-Department of Health and Social Care: (i) "hands," proper and systematic hand cleaning; (ii) "face," respiratory protection via masks or face coverings; (iii) "space," minimum spatial distance between individuals in case of contact.

---

*The chapter is the result of joint research carried out by the three authors. More specifically, Fulvio Adobati compiled Sections 6.1.1 and 6.1.5; Emanuele Comi Sections 6.1.2.1, 6.1.2.2, and 6.1.3; Alessandra Ghisalberti Sections 6.1.2.3 and 6.1.4.

6.1.2 Contagion containment measures in Europe and their space–time evolution

6.1.2.1 The measures taken by the European Union

It is well known that among its constituent principles, the European Union includes the free movement of people and goods across Member States.[a] The imposition of limits set by Member States on freedom of movement on account of Covid-19 contagion and the prediction, on the part of EU agencies, of further freedom restrictions attest to the gravity of the epidemic.

One of the first, albeit nonprescriptive, official proceedings whereby a position was taken on the measures adopted was European Commission Communication COM (2020) no. 115 of 16 March 2020. At a juncture when some countries, such as Italy, had already foreseen a general closure of activities and stringent travel restrictions, the Communication seems to legitimize measures independently adopted by individual countries. Generalized restrictions on circulation are recommended: however, with a view to ensuring policy coordination, exceptions to be warranted are also indicated.[b] The adoption of coordinated and binding rulings for Member States is ultimately deferred to the European Council. Also on March 16, 2020, in the Official Journal of the European Union, the "Guidelines on border management measures designed to protect health and ensure the availability of essential goods and services" were issued by the European Commission.[c] While these have no mandatory scope, they do mark an attempt to encourage the adoption of measures for restricting movement while safeguarding the free movement of goods within the Union.

From March 17, 2020, following the aforementioned Commission proceedings, Member States have begun to introduce generalized forms of restriction to circulation in a coordinated approach.[d]

Subsequently, on March 30, 2020, with Communication C (2020) no. 2050, the Commission adopted guidelines, still of a nonbinding nature, on the implementation of temporary restrictions for nonessential travel within the European Union and, again on March 30, with Act no. 2020/C 102 I/03, of guidelines on the circulation of workers. With Communication COM (2020) no. 148 of April 8, 2020, an extension of measures beyond the monthly timeframe envisaged at the beginning of March was legitimized.

Once again via nonbinding Acts, on April 8, 2020 and April 16, 2020 the Commission set out recommendations and guidelines on infection tracing devices.[e]

---

[a] According to the European Court of Justice, "Treaty articles related to the free movement of goods, persons, services and capital constitute fundamental Community rules and [...] any obstacle to such freedom, albeit minor, is forbidden (in particular, see sentences dated 13 December 1989, case C-49/89, Corsica Ferries France, ECR 4441, paragraph 8, and 15 February 2000, case C-169/98, Commission v France, ECR p. I-1049, paragraph 46)" (CJUE, sentence 1 April 2008, in case C/212-06, par. 52). Literature on this point is vast. By way of example see: Rossi Dal Pozzo (2012), Daniele (1989).

[b] Such as, for instance, the right of Union citizens, residents, and their families to return home; the movement of workers engaged in essential services and the rights of passengers traveling for "imperative family reasons."

[c] Act no. 2020/C 86 I/01.

[d] This is reflected in Communication of the European Commission COM (2020) no. 148 of 8 April 2020.

[e] EU Commission, Recommendation (EU) 2020/518 of 8 April 2020 and Communication 2020/C 124 I/01 no. C/2020/2523 of 16 April 2020.

On 15 April, however, the European Commission and the President of the European Council adopted a "roadmap" to encourage plans for a gradual reopening of borders, and the lifting of containment measures.[f] Again, on May 8, 2020, faced with the persistence of the epidemic crisis—the Commission recommended that Member States extend temporary restrictions on non-essential travel until 15 June.[g] Nonetheless, with a series of Acts of May 13, 2020, it also issued guidelines in view of hopeful reopenings.[h]

Finally, due to the slowdown of the epidemic, with recommendation dated June 11, 2020, the Commission called for the lifting of internal controls at the Union's borders, keeping in place only travel restrictions outside Europe.[i]

At the end of this brief overview, it should be acknowledged that, by adopting the measures mentioned above, the European Union did legitimize epidemic containment policies in the form of general restrictions on freedom of movement. At the same time, however, the Union aimed from the start to guarantee support to States most affected by the emergency, for instance by facilitating the transfer of patients to less affected countries.[j] Above all, it strove (via the common market) to ensure market availability of medical devices,[k] to the point of waiving taxation on such products in order to contain prices.[l]

6.1.2.2 Measures adopted by the European States and measure severity index

As mentioned above, since 16 and 17 March 2020, in accordance with (albeit nonbinding) guidelines laid out by the Commission, Union States together with other non-Union European States, have progressively introduced generalized and coordinated forms of circulation restriction.[m]

European Commission guidelines were implemented by single States in a variegated range of different measures, tied to local epidemiological contingencies, to the nature of each

---

[f]More information of this roadmap may be found at: https://ec.europa.eu/info/files/communication-european-roadmap-lifting-coronavirus-containment-measures_en.

[g]European Commission, Communication COM (2020) 222 of 08 May 2020.

[h]European Commission, Communication 2020/C no. 169/03 C/2020/3250 of 13 May 2020; Communication COM (2020) 550 of 13 May 2020.

[i]European Commission, Communication COM (2020) 399 of 11 June 2020.

[j]By supporting cooperation in cross-border healthcare since European Commission Communication C (2020) no. 2153 of 03 April 2020 on the subject of "Guidelines on EU Emergency Assistance in Cross-Border Cooperation in Healthcare related to the Covid-19 crisis."

[k]For instance, by authorizing export outside the Union of personal protective equipment via Implementing Regulation (EU) 2020/402 of 14 March 2020; yet—by contrast—stigmatizing prohibitions on supply to other EU countries (for example, in the aforementioned COM (2020) no. 115 of 16 March 2020) or public hoarding in the absence of valid health reasons (see Commission Communication 2020/C no. 116 I/01 of 08 April 2020). Also, by providing "green lines" along which border controls could be carried out swiftly to guarantee supplies (Commission Communication C (2020) no. 1897 of 23 March 2020).

[l]Commission Decision C (2020) no. 2146 of 3 April 2020 which, for 6 months, exempted face masks, protective devices, testing kits, ventilators and other medical equipment from VAT and customs duties.

[m]This is reflected in Communication of the European Commission COM (2020) (2020) no. 148 of 8 April 2020.

country's constitutional framework and, finally, to political choices made by legislative and executive bodies.

Some European academic institutions sifted the vast amount of data available on the measures taken by individual States: they have conducted studies on such measures and, in some cases, on their tangible impact. In particular, research conducted at the London Imperial College on the impact of nonmedical-pharmaceutical measures on viral spread (Flaxman et al., 2020a) have pointed out that—for the rest of Europe, with the exception of Italy, which had already adopted initial containment measures—States on the European continent began to place restrictions exactly between 16 and 17 March 2020 (Flaxman et al., 2020a, p. 5). The study carried out by the Blavatnik School of Government at the University of Oxford reached even farther. It examined measures adopted by all States and developed a severity, or *stringency index* which outlines the pervasiveness of the measures adopted by the State, as also shown in the following figure.

6.1.2.3 A severity index map for Europe

The "stringency index" developed at the University of Oxford quantifies the severity level of political measures deployed by individual countries worldwide in response to the Covid-19 epidemic.[n] This index was computed by cross-referencing 17 different indicators related to policies activated by each government for Covid-19 containment across various areas, such as restrictions on individual mobility or closure of production and commercial activities. An index evolution analysis makes it possible to carry out a comparative study on Covid-19 containment interventions by individual States in a space–time perspective, using interactive mapping devised by the same Oxford research group for visualizing the evolution of severity of policies starting from January 21, 2020 to the present.

Fig. 6.1 shows select images from the "Oxford Covid-19 Government Response Tracker" mapping system, which enable us to track the "stringency index" evolution over time and to record Italy's swift political intervention, a precursor with regard to the wider European territory.

Specifically, Fig. 6.1 shows the severity index in European countries on four different dates between February and June 2020—the period covered by our study. These account for Italy's pioneering role in epidemic containment policies. Quadrant (a) in the figure, referred to February 23, 2020, is color-coded in dark blue to mark the highest level of severity—solely recorded in Italy—of Covid-19 containment policies at the onset of the epidemic phase. National Statutory order (D.L.N.) 6—to be addressed in detail in the pages that follow—was in fact approved on that date. It set out urgent measures (unprecedented in Europe) for the containment and management of the Covid-19 epidemiological emergency across the Italian territory, and established a so-called "red zone" in the area of the two Municipalities of Codogno and Vo' Euganeo. The stringency of Italian policies at the time is in sharp

[n]We are thinking here of research outcomes achieved at the Blavatnik School of Government of Oxford University, which produced the "Oxford Covid-19 Government Response Tracker" (OxCGRT) system for collecting and displaying information on the policies adopted by 180 countries around the world. In particular, the research team produced interactive and diachronic mapping which may be consulted here: https://www.bsg.ox.ac.uk/research/research-projects/coronavirus-government-response-tracker.

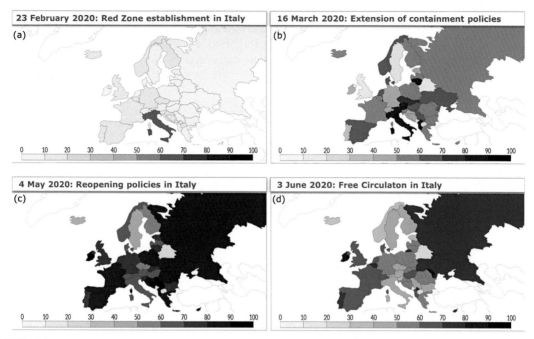

FIG. 6.1 Different containment measures in European Countries between February and June 2020. Source: Local processing of "Oxford Covid-19 Government Response Tracker" data, University of Oxford, 2020.

contrast with other European countries where no rigid policies had yet been implemented, a fact visualized by light-yellow color-coding on the map. Conversely, quadrant (b) records a subsequent increase in the stringency index for various other European countries and, as of March 16, 2020, an extension of containment policies in much of Europe, here color-coded in light blue. Even on that date, however, the marked stringency of containment policies in Italy, most affected by the epidemic, is evident and color-coded in dark blue. As the infection reaches its endemic phase, on May 4, 2020 (c), a consistent trend reversal may be seen across Italy, with the introduction of reopening policies. Color-coding in this case is seen to fade for the first time on Italy, while it takes on a darker shade across most of the European territory. Finally, quadrant (d) shows a further attenuation of the stringency index of Italian policies coinciding with the reopening of free circulation in Italy starting June 3.

Ultimately, digital mapping makes it possible to compare the evolution of severity level in containment policies adopted by individual European countries over time and promotes a spatio-temporal interpretation of the phenomenon. In particular, on the one hand, it underlines the difficulty of coordinated and consistent political intervention in Europe, to be ascribed for example to the absence of a common strategy for combating contagion, to unilateral and uncoordinated border closures, or to incongruous lockdown measures (Lumet and Enaudeau, 2020). On the other hand, it highlights the swift political intervention of the Italian Government in the face of an exponential rise of infection during the epidemic phase, first, and its equally swift containment in the endemic phase, later.

Cartography highlights the pioneering role of Italy in the European context as a reference model in emergency management for various European countries. At the same time, digital mapping also shows that rapid intervention in Italy had a positive impact on the evolution of contagion because it facilitated containment, as discussed in the pages that follow.

6.1.3 The Italian legal system

6.1.3.1 Introduction

As we turn to the Italian context, we may note that Italian public administrations have played a significant role in combating the Covid-19 epidemic and in managing the health emergency that ensued.[o] Associations and private individuals also played a role in these areas, participating in many ways with independent initiative or via action coordinated with the public administration.[p] Initially, however, a range of actions, for instance serological testing, were set up exclusively via public monopoly and excluded private sector agencies.[q]

Due to the scope of research, we will focus exclusively on the role played by public administrations and on the policies undertaken that may have affected the handling and containment of the epidemic.

In a nutshell, in the management of emergencies, and notably of health emergencies, the Italian legal system provides for a range of competitive legislative sources. Excluding legislative sources, and especially urgency provisions,[r] there exists in the first place a government-based administrative jurisdiction, which may be exercised by the President of the Council of Ministers,[s] by the Head of Civil Protection[t] and by the Minister of Health.[u] These legislative and provisionary powers, entrusted to government bodies, are matched by the competence of Region Governors[v] and Mayors.[w] A distinctive feature of such powers is that they are on the whole atypical. In other words, they may address any type of content and, accordingly, prescribe any type of writ.

The analysis will focus mainly on the Lombardy region and on Lombard local administrations, as well as on bordering regions which faced the most severe effects of the first epidemic wave.

[o] It seems crucial, in fact, to underline the eminently public role of collective health protection.

[p] For example, donation-based building of hospitals, PPE production or purchase of medical equipment.

[q] In Lombardy, DGR (Regional Government Decree) 12 May 2020, no. XI/3131 eventually established that private testing facilities may offer services of this kind, even outside the conventional policies of the regional healthcare system.

[r] Albeit frequently used to deal with emergency, both to provide for specific measures where required by law, and to waive other ordinary legislative provisions.

[s] TU civil protection.

[t] TU civil protection.

[u] Art. 32 SSN-National Health System institutive decree.

[v] Art. 32 SSN-National Health System institutive decree.

[w] Art. 50 of Legislative Decree no. 267/2000.

6.1.3.2 National containment measures in Italy

The first formal Act marking a response to the Covid-19 health emergency was the declaration of the state of emergency approved by the Council of Ministers on January 31, 2020.[x] The provision falls within the emergency powers permanently endowed to the government by civil-protection legislation and, in particular, by the so-called "Civil protection code," adopted by legislative decree 2 January 2018, no. 1[y] and substantially amended at the very onset of emergency by legislative decree 6 February 2020, no. 4.

Possibly faced with the inadequacy of ordinary emergency management tools to deal with the Covid-19 epidemic, the Government—in the exercise of its emergency legislative powers set forth by art. 77 of the Constitution—approved legislative decree February 23, 2020, no. 6. With this statutory order, an unprecedented method of exercising emergency powers was outlined in the Italian system, at least in the humble opinion of the present writer. The decree in fact endows the President of the Council of Minister with a very broad range of powers for restricting constitutionally guaranteed rights.[z]

At the same time, the Legislative Decree 6/2020 seeks to outline a redefinition of ordinance powers in healthcare matters historically ascribed to the multiple administrative agencies mentioned above.[aa]

Via a DPCM also dated February 23, the President of the Council of Ministers began to adopt—among others—a series of prescriptions restricting freedom of movement and business activity. These were subsequently tightened up through the same administrative instrument as the situation worsened and finally alleviated as normality was progressively reestablished. At the same time, both national ordinances (of the Civil Protection and the Minister of Health) and regional and municipal ordinances were brought in order to issue alternative regulations on what already established by national legislation or to introduce further provisions, both towards a restriction or extension of freedoms.

6.1.3.2.1 Restrictions on freedom of movement

As mentioned, the first Act of actual restriction to freedom of movement was by Decree of the President of the Council of Ministers (DPCM) of 23 February 2020, whereby mobility to (Article 1, paragraph 1, letter a) and from (art 1, paragraph 1, letter b) the municipalities of the Codogno area[ab] in Lombardy and Vo' Euganeo in Veneto was forbidden, and a corresponding red zone" was established.

[x]The provision is entitled "Declaration of a state of emergency as a consequence of the health risk associated with the onset of pathologies deriving from transmissible viral agents."

[y]"Ordinary" emergency legislation has been in place since 1992. Legislative Decree no. 1/2018 was heavily amended by legislative decree no. 4/2020 of February 6, 2020 with regards to areas of competence.

[z]This approach is confirmed by subsequent law decrees, which generally and abstractly impose a very wide range of restricting powers onto the legal sphere of subjects, the actual exercise of which is delegated to the Decree of the President of the council of Ministers.

[aa]Restriction of local authorities to rules "not in contrast" with those of the State.

[ab]Municipalities included (a) Bertonico; (b) Casalpusterlengo; (c) Castelgerundo; (d) Castiglione D'Adda; (e) Codogno; (f) Fombio; (g) Maleo; (h) San Fiorano; (i) Somaglia; (j) Terranova dei Passerini.

DPCM dated March 1, 2020 confirmed the ban on mobility from the municipalities of the so-called red zone of Codogno and of Vo' Euganeo (Article 1, paragraph 1, letter a) and renewed the prohibition to access the same municipalities (Article 1, paragraph 1, letter b).

As of 8 March, for Lombardy and for other Italian provinces where infection rates were high,[ac] the decree of the President of the Council of Ministers urged citizens to "avoid" any movement into and out of the territories or within the same territories.[ad] This recommendation was extended to the whole national territory on March 9, 2020,[ae] and included a ban on gatherings.[af]

As of March 22, 2020, citizens were thoroughly banned from moving to a municipality other than the one in which they resided, except for proven work needs, of absolute urgency or for health reasons.[ag]

Measures were partially lifted on April 26, when the meeting of relatives was admitted, albeit without going beyond regional boundaries.[ah] As of the same date, those who had found themselves outside their own municipality on March 22 were also allowed back home.

Finally, as of June 3, provisions restricting travel between regions ceased to be enforced, effectively reopening circulation within the country.

6.1.3.2.2 *Restrictions on educational, community and collective recreational activities*

Still in the Codogno area in Lombardy and Vo' Euganeo in Veneto, starting as for February 23, 2020, all events or initiatives of any nature and any form of meeting in either public or private venues were suspended (Article 1, paragraph 1, letter c), as were education services and schools of any rank or level (Article 1, paragraph 1, letter d) and school trips (Article 1, paragraph 1, letter e). Again, as of the same date areas museums and similar venues were closed to the public across the same area (art. 1, paragraph 1, letter f).

In all municipalities for the regions of Emilia Romagna, Friuli Venezia Giulia, Lombardy, Veneto, Liguria and Piedmont, school trips and educational visit were suspended as of 25 February.[ai]

As of March 1, 2020, the suspension of school trips and educational visits was extended to the whole of Italy (Article 4, paragraph 1, letter b).

Also, as of 1 March 2020 in the regions of Emilia-Romagna, Lombardy, and Veneto and for the provinces of Pesaro-Urbino and Savona, suspension of events or initiatives of any nature, of events and of any form of meeting in public venues or in nonordinary private venues (Article 2, paragraph 1, letter c) was enforced. Similarly, in the same area, suspension of all education services and schools of any rank or level was decreed (Article 2, paragraph

---

[ac] The provinces of: Modena, Parma, Piacenza, Reggio Emilia, Rimini, Pesaro-Urbino, Alessandria, Asti, Novara, Verbano-Cusio-Ossola, Vercelli, Padua, Treviso and Venice.

[ad] DPCM 08/03/2020, art. 1, lett. a.

[ae] DPCM 09/03/2020, art. 1, no. 1.

[af] DPCM 09/03/2020, art. 1, no. 2.

[ag] DPCM 22 March 2020, art. 1, lett. b.

[ah] DPCM 26 April 2020, art. 1, lett. a.

[ai] DPCM February 25, 2020, art. 1, paragraph 1, lett. b.

1, letter e). As of the same date, openings of museums and cultural venues in the aforementioned regions and provinces was permitted, provided social distancing of at least 1 m could be ensured (Article 2, paragraph 1, letter f).

As of 4 March, all conference and congress activities were suspended throughout the national territory (Article 1, paragraph 1, letter a), as were events and shows held in public or private venues that failed to comply with 1-m social distancing rules (Article 1, paragraph 1, letter b). Finally, as of 8 March this suspension measure was extended throughout, regardless of interpersonal distances (Article 2, letter a).

As of the day after March 4, all education services and teaching activities were suspended throughout the country (Article 1, paragraph 1, letter d).

As of 8 March, suspension of public events or events in private venues was also enforced, and extended to "ordinary" public events, such as theatrical performances, cinemas, pubs, gaming and betting rooms, dance schools and dance clubs, which were enjoined to cease all activity.[aj] Museums were also closed to the public.[ak]

As of May 17, public events in the open in static form were once again allowed—in compliance with social distancing.[al] Also, as of May 17, museums were allowed to reopen, with prescriptions regarding social distancing between visitors.[am]

In the same Act, reopening of theatrical and cinema halls is envisaged. It will be allowed as of June 15, 2020.[an]

6.1.3.2.3 *Restriction of sports activities*

As of February 25, 2020, for all municipalities of the Regions of Emilia-Romagna, Friuli-Venezia Giulia, Lombardy, Veneto, Liguria, and Piedmont all sporting events involving a public audience were suspended, with the sole permission to hold events and training behind closed doors.[ao] This exception, however, did not apply to the so-called red zones of Codogno and Vo' Euganeo, for which sporting events or training were banned even behind closed doors.[ap]

As of 1 March, suspension of "open door" sporting events was confirmed only for Emilia-Romagna, Lombardy, and Veneto and for the provinces of Pesaro-Urbino, and Savona. Permission to hold events and training behind closed doors was also confirmed, with the exception of red zones.[aq] Also as of March 1, 2020, sport supporters residing in the regions

---

[aj]DPCM 08 March 2020, art. 1, lett. G for Lombardy and other provinces, art. 2, lett. c for the entire national territory.

[ak]DPCM 08 March 2020, art. 1, lett. G for Lombardy and other provinces, art. 2, lett. d for the entire national territory.

[al]DPCM 17 May 2020, art. 1, lett. i.

[am]DPCM 17 May 2020, art. 1, lett. p.

[an]DPCM 17 May 2020, art. 1, lett. m.

[ao]DPCM February 25, 2020, art. 1, paragraph 1, lett. a.

[ap]DPCM February 25, 2020, art. 1, paragraph 1, lett. a.

[aq]DPCM 01 March, 2020, art. 1, paragraph 1, lett. a.

of Emilia-Romagna, Lombardy, and Veneto and in the provinces of Pesaro-Urbino and Savona were banned from traveling to attend "open door" sporting events held in other areas.[ar] As of the same date ski areas activities were allowed, but with closed transport restrictions aimed at preventing crowding.[as]

Also as of 1 March in Lombardy and in the province of Piacenza, all activity for gyms, sports centers and swimming pools was suspended (Article 2, paragraph 3, letter a). As of 8 March this closure is also extended to the provinces of Modena, Parma, Piacenza, Reggio nell'Emilia, Rimini, Pesaro-Urbino, Alessandria, Asti, Novara, Verbano-Cusio-Ossola, Vercelli, Padua, Treviso and Venice.[at]

Throughout the national territory, as of 4 March 2020, "open door" sports events and competitions were canceled, with events and training sessions without an attending public were still permitted (Article 1, paragraph 1, letter c). The activity of gyms, swimming pools and sports centers was allowed provided compliance was ensured with 1-m social distancing rules (Article 1, paragraph 1, letter c).[au] Events and trainings were suspended throughout Italy as of 1 April 2020.[av]

As of March 8, 2020, in Lombardy and in provinces with the highest infection rates,[aw] ski facilities in skiing areas were closed.[ax]

The lifting of measures started on April 26, with the reopening of public parks and gardens and with the permission to carry out sports activities in compliance with social distancing.[ay] Competitive training at sports facilities was also allowed to resume.[az]

Finally, as of 12 June, sporting competitions were allowed to resume behind closed doors throughout Italy.[ba]

6.1.3.2.4 Restrictions to commercial and productive activities

Once again with regard to the Codogno area in Lombardy and Vo' Euganeo in Veneto, as of 23 February 2020, the following measures were put in place: closure of all commercial activities, with the exception of those for the purchase of basic necessities, public utility or essential public services (Article 1, paragraph 1, letter i); suspension of freight and passenger transport services, including non-scheduled services (Article 1, paragraph 1, letter m); suspension of work activities in the above-mentioned area, with the exception of those providing essential or public utility services and those that may be carried out at home or remotely (Article 1,

---

[ar] DPCM 01 March, 2020, art. 2, paragraph 1, lett. a.

[as] DPCM 01 March, 2020, art. 2, paragraph 1, lett. b.

[at] DPCM 08 March 2020, art. 1, lett. s.

[au] Such prescription was reaffirmed as of DPCM 08/03/2020, art. 1, lett. d.

[av] DPCM 01 April 2020, art. 1, no. 2.

[aw] The provinces of: Modena, Parma, Piacenza, Reggio Emilia, Rimini, Pesaro-Urbino, Alessandria, Asti, Novara, Verbano-Cusio-Ossola, Vercelli, Padua, Treviso and Venice.

[ax] DPCM 08/03/2020, art. 1, lett. f. The reopening of ski areas would only take place on 11 June 2020.

[ay] DPCM 26 April 2020, art. 1, lett. e and f.

[az] DPCM 26 April 2020, art. 1, lett. g.

[ba] DPCM 11 June 2020, art. 1, lett. e.

paragraph 1, letter n) and, finally, suspension of work activities for workers who reside inside the affected area but work outside the red zone (Article 1, paragraph 1, letter o).

As of March 1, 2020 in the regions of Emilia-Romagna, Lombardy, and Veneto and for the provinces of Pesaro-Urbino and Savona, catering services, bar and pub activities were permitted only for seated customers and in compliance with social distancing rules (Article 2, paragraph 1, letter h).

Also, as of March 1, the remaining commercial activities were enjoined to ensure safe social distancing between visitors (Article 2, paragraph 1, letter i). As of 8 March this prescription was extended to provinces with high infections rates,[bb] while it took the form of a recommendation for the rest of Italy,[bc] but was made mandatory for the rest of the country on March 9.

As of March 1, only for the provinces of Bergamo, Lodi, Piacenza, and Cremona, complete shutdown of medium and large shops, commercial establishments inside shopping malls or markets was ordered on Saturdays and Sundays, with the exception of pharmacies and drugstores, as well as grocery stores. As of March 8, this type of shutdown was also extended to the provinces of Modena, Parma, Piacenza, Reggio Emilia, Rimini, Pesaro-Urbino, Alessandria, Asti, Novara, Verbano-Cusio-Ossola, Vercelli, Padua, Treviso, and Venice.[bd]

As of March 8, 2020 in Lombardy and in the provinces with high rates of infection[be] restaurant and bar activity was permitted only from 6 am to 6 pm,[bf] while for the rest of Italy safe social distancing must be ensured.[bg] As of March 9 restrictions were extended to all of Italy.

As of March 11, all nonfood businesses or markets, even outdoors, which did not provide basic necessities were closed.[bh] As of the same date, bars and restaurant businesses are also shut down, with the exception of home delivery.[bi] Hairdressers and beauty parlors were also suspended.[bj]

A further squeeze occurs on March 22, with the closure of all industrial and commercial production activities, with the sole exclusion of activities deemed relevant to address the emergency[bk] and of the activities instrumental to the latter, subject to prior communication by the Prefecture.[bl]

Reopening of nonfood commercial activities and catering services, as well as of personal care services (hairdressers and beauticians) took place from May 17, 2020.

Finally, on June 11, gaming and betting rooms reopened, unless otherwise specified by the Region.

[bb] DPCM 08 March 2020, art. 1, lett. o.

[bc] DPCM 08 March 2020, art. 2, lett. f.

[bd] DPCM 08 March 2020, art. 1, lett. r.

[be] The aforementioned provinces of: Modena, Parma, Piacenza, Reggio Emilia, Rimini, Pesaro e Urbino, Alessandria, Asti, Novara, Verbano-Cusio-Ossola, Vercelli, Padua, Treviso and Venice.

[bf] DPCM 08 March 2020, art. 1, lett. n.

[bg] DPCM 08 March 2020, art. 2, lett. e.

[bh] DPCM 11 March 2020, art. 1, no. 1.

[bi] DPCM 11 March 2020, art. 1, no. 2.

[bj] DPCM 11 March 2020, art. 1, no. 3.

[bk] DPCM 22 March 2020, art. 1, lett. a.

[bl] DPCM 22 March 2020, art. 1, lett. g.

6.1.3.2.5 Restrictions on worship activities

Always in the Codogno area in Lombardy and Vo' Euganeo in Veneto, as of February 23, 2020, suspension of religious services was mandated (Article 1, paragraph 1, letter c).

As of 1 March in the regions of Emilia-Romagna, Lombardy, and Veneto and for the provinces of Pesaro, Urbino and Savona, suspension of religious services was ordered (Article 2, paragraph 1, letter c). The opening of places of worship was permitted provided compliance to 1-m safe social distancing measures could be ensured (Article 2, paragraph 1, letter d). As of 8 March such prescriptions were extended not only to provinces with high infection rates,[bm] but also to the rest of Italy.[bn]

As of April 26, 2020, gradual reopening only allows the celebration of funeral services with a maximum participation of 15 relatives.[bo]

Finally, as of May 17, 2020 permission for public worship in compliance with social distancing measures was reinstated.[bp]

6.1.3.2.6 Promotion of flexible work

As of February 23, 2020, strategies were put in place to promote work from home via available online technology. Art. 3 of the Decree of the President of the Council of Ministers (DPCM) dated February 23, 2020 mandated that every current employment relationship in the so-called red zone should provide for flexible work, regulated by articles 18 and ff. of 23 law 22 May 2017, no. 81, even in the absence of specific agreements.

This opportunity was later extended from February 25, 2020 to all employers based in the regions of Emilia-Romagna, Friuli-Venezia Giulia, Lombardy, Piedmont, Veneto, and Liguria, as well as to workers residing in the same area who carried out work activities outside these territories.[bq]

As of March 1, 2020, permission to resort to flexible work was extended to the whole national territory (Article 4, paragraph 1 letter a).

Also as of March 1, 2020, for the regions of Emilia-Romagna, Lombardy, and Veneto and for the provinces of Pesaro-Urbino and Savona recommendations were issued with regard to favoring remote connection "in the conduct of meetings or conferences" (Article 2, paragraph 1, letter m). Such recommendation was reiterated throughout the national territory on March 8, 2020.

6.1.3.2.7 Public offices and certificates

Always in the Codogno area in Lombardy and Vo' Euganeo in Veneto, as of February 23, 2020, all public office activities were banned, without prejudice to the provision of essential and public utility services (Article 1, paragraph 1, letter g). As of February 25, 2020, judicial

---

[bm] DPCM 8 March 2020 art. 1, lett. i, Modena, Parma, Piacenza, Reggio Emilia, Rimini, Pesaro-Urbino, Alessandria, Asti, Novara, Verbano-Cusio-Ossola, Vercelli, Padua, Treviso, and Venice.

[bn] DPCM 08 March 2020, art. 2, lett. V.

[bo] DPCM 26 April 2020, art. 1, lett. I.

[bp] DPCM 17 May 2020.

[bq] Art. 2 DPCM February 25, 2020.

offices to which the municipalities in question belong were generally closed to the public except for urgent certificates. Reduced schedules were arranged.[br]

As of February 25, 2020, in the provinces of Bergamo, Brescia, Cremona, Lodi, Milan, Padua, Parma, Pavia, Piacenza, Rovigo, Treviso, Venice, Verona and Vicenza, driving tests and access to civil motorization offices were suspended, with the provision of an extension of expiration terms for some documents.[bs] Such extension was subsequently granted to other areas.[bt]

As of March 1, 2020 in the regions of Emilia-Romagna, Lombardy, and Veneto and for the provinces of Pesaro-Urbino and Savona, public and private competitive exams requiring the presence of candidates were suspended (Article 2, paragraph 1, letter g), while for the remaining areas of Italy the need to enforce safe social distancing was asserted (Article 3, paragraph 1, letter f). Such suspension was extended on 8 March also to provinces with high infection rates.[bu] As of 9 March 2020, a general suspension was extended to all of Italy.

6.1.3.2.8 *Obligations to notify travel*

As of 1 February 2020, those who passed through or stopped within the Codogno area in Lombardy and Vo' Euganeo in Veneto were obliged to notify the competent Local Health Authority (ASL) Prevention Department "for the purpose of adoption, by the competent health authority, of all necessary measures, including active-surveillance home stay."[bv]

By Decree of the President of the Council of Ministers (DPCM) dated March 1, 2020, those who entered Italy from epidemic risk areas in the previous 14 days were bound to notify the competent Local Health Authority Prevention Department in order to possibly arrange for health surveillance and fiduciary isolation (Article 3, paragraph 1, letter g and Article 3, paragraphs 2–6).

6.1.3.2.9 *Obligation to wear personal protective equipment*

As of February 23, 2020, those based in the Codogno area in Lombardy and Vo' Euganeo in Veneto were the first to be required to wear personal protective equipment (or to adopt specific precautionary measures set out by the Local Health Authority Prevention Department) to access essential public services, as well as commercial establishments (Article 1, paragraph 1, letter l).

[br] DPCM 25 February 2020, art. 1, paragraph 1, lett. l. Such provision was confirmed by art. 2, paragraph 4 of DCPM 01 March 2020.

[bs] DPCM 25 February 2020, art. 1, paragraph 1, lett. f and g.

[bt] DPCM 01 March 2020, art. 1, paragraph 1, lett. o.

[bu] Modena, Parma, Piacenza, Reggio Emilia, Rimini, Pesaro-Urbino, Alessandria, Asti, Novara, Verbano-Cusio-Ossola, Vercelli, Padua, Treviso and Venice.

[bv] DPCM February 23, 2020, art. 2.

6.1.3.3 Measures applied to those who were already deprived of personal freedom

With regard to those expected to go to prison or to minor correctional facilities, especially if coming from the Codogno area in Lombardy or Vo' Euganeo in Veneto, a health protocol was to be set up as of February 25, 2020.[bw]

The need to limit contagion in prisons or similar facilities by implementing adequate safeguards and enforcing protocols was repeatedly affirmed[bx] even later.

In the regions of Emilia-Romagna, Lombardy, and Veneto and for the provinces of Pesaro-Urbino and Savona, Decree of the President of the Council of Ministers (DPCM) dated 1 March 2020 called for "severe restriction" of visitor access to guests in Nursing and Residential Care Facilities.

As of 4 March, throughout the national territory, access of relatives and visitors to family members in the hospitality and long-term hospitalization facilities, in assisted healthcare residences or residential facilities for the elderly could be restricted by the facility's health management (Article 1, paragraph 1, letter m).

Finally, as of March 1, 2020, in the regions of Emilia-Romagna, Lombardy and Veneto and for the provinces of Pesaro-Urbino and Savona, hospital health departments were asked to restrict visitor access to hospitalization areas (Article 1, paragraph 1, letter j).

As of March 4, 2020, throughout the national territory, it was expressly forbidden for people accompanying patients to wait in emergency rooms (Article 1, paragraph 1, letter l).

6.1.3.4 Regional and local measures to contain infection

As mentioned above, management of health emergencies generally also entails a concurrent ruling power on the part of Region Governors[by] and Mayors of municipalities.[bz]

In emergency legislation proper to the onset phase of the epidemic, the legislator endeavored to rank the various administrative sources of production of contagion containment provisions, assigning a prominent role to the decrees of the President of the Council of Ministers. The art. 3, paragraph 1 of the Legislative Decree February 23, 2020, no. 6 entrusts the implementation of containment measures envisaged to the President of the Council of Ministers following a summary participatory procedure.[ca] Only "pending adoption" of Decrees of the President of the Council of Ministers (DPCMs), in cases of "extreme necessity and urgency," can the bodies of local authorities adopt the ordinances mentioned.[cb]

[bw] DPCM February 25, 2020, art. 1, paragraph 1, lett. m.

[bx] DPCM 04 March 2020, art. 1, paragraph 1, lett. p.

[by] Art. 32 of Law 23 December 1978, no. 833.

[bz] Art. 50 of Legislative Decree no. 267/2000.

[ca] In fact, the decree may be adopted "upon proposal of the Minister of Health, after consulting the Minister of the Interior, the Minister of Defense, the Minister of Economy and Finance and other Ministers competent for the matter, as well as the Presidents of competent Regions, in the event that they exclusively concern a single region or some specific regions, or the President of the Conference of Region Presidents, in the event that they concern the national territory" (art. 3, paragraph 1 of Legislative Decree February 23, 2020, no. 6).

[cb] Art. 3, paragraph 2 of Legislative Decree 23 February 2020, no. 6.

Following the adoption of a flood of local ordinances, the legislator, by Legislative Decree March 2, 2020, no. 9 urgently established that "following the adoption of State measures to contain and manage emergency […], contingent and exceptionable ordinances aimed at dealing with the aforementioned emergency may not be implemented and, where implemented ought to be considered invalid as against State measures."[cc]

This requirement was reaffirmed by Legislative Decree March 25, 2020, no. 19 which sanctioned regional competence temporally, solely "pending the adoption of decrees of the President of the Council of Ministers […] and with limited effectiveness until that moment" and exclusively "in relation to specific supervening situations of aggravation of the health risk that occurred in their territory or part of it."[cd] In such cases, Regions "may introduce further restrictive measures […] exclusively in the context of the activities within their competence and without affecting production activities and those of strategic importance for the national economy."[ce] Even more restricted is the competence area of Mayors, since a municipal body may not adopt, "under penalty of invalidity, extraordinary and urgent ordinances aimed at facing an emergency in contrast with State measures, nor exceeding the limits" provided for State regulations.[cf]

It was asked how one could discern, in this specific context, the "contrast" between local ordinances and State provisions: in a nutshell, administrative law has—in general terms—acknowledged the legitimacy of local ordinances which were more restrictive than national ordinances, with limited exceptions for notably sensitive issues of exclusive state competence.[cg]

6.1.4 Italian measures in a space–time perspective

Overall, Covid-19 contagion containment measures in Italy may be traced back to a space–time interpretative model, taking into account the three phases whereby the epidemic spread and their impact on territory.[ch] In fact, as shown in the Fig. 6.2, political containment measures chart an evolution over time which reflects the three phases of contagion spread: the onset, epidemic and endemic phases.

The first phase—the onset phase—emerged when definite and full-blown cases of Covid-19 contagion were recorded. That lead the Italian government to declare a state of emergency by the Legislative Decree no. 4 of February 6, 2020. As marked on the map by

[cc] Art. 35 of Legislative Decree 8 March 2020, no. 9.

[cd] Art. 3, paragraph 1 of Legislative Decree 25 March 2020, no. 19.

[ce] Art. 3, paragraph 1 of Legislative Decree 25 March 2020, no. 19.

[cf] Art. 3, paragraph 2 of Legislative Decree 25 March 2020, no. 19.

[cg] As in the case of immigration or the lockdown of means of transport.

[ch] We refer here to the theoretical-methodological approach outlined by Emanuela Casti in the introduction to this volume, which breaks down the evolution of Covid-19 infection in Italy into three phases, namely: the onset phase, when first cases, either suspect or overt, are recorded; the epidemic phase, when infection spreads rapidly throughout the entire community; the endemic phase, when the number of infected individuals decreases but does not disappear. For a full treatment of this issue, see also: Casti (2020), pp. 65–66.

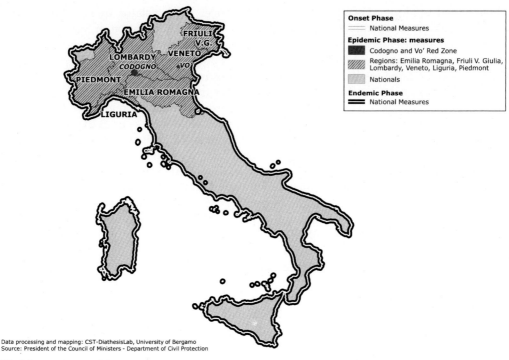

Data processing and mapping: CST-DiathesisLab, University of Bergamo
Source: President of the Council of Ministers - Department of Civil Protection

FIG. 6.2 Evolution of the containment measures in Italy in February and March 2020.

a white line, such measure affected the entire Italian territory and followed the initial onset of Covid-19.

The second phase—the epidemic phase—ensued as the spread of contagion grew exponentially and triggered restrictions which applied to different territorial contexts: definition of the so-called "red zone," following DPCM February 23, 2020, in the Municipalities of the Codogno area[ci] in Lombardy and in the Municipality of Vo' Euganeo in Veneto, color-coded in gray lines on the map, which banned movement into or from the area, preventing people's mobility; restriction of educational, community and collective recreational activities via DPCM February 23, 2020 for the Regions of Emilia-Romagna, Friuli-Venezia Giulia, Liguria, Lombardy, Piedmont and Veneto, color-coded in red on the map; finally, extension of restrictions to the whole Italian territory—color-coded in light gray on the basemap—via multiple national provisions issued in close succession starting with the Decree of the President of the Council of Ministers dated March 1, 2020.

Finally, following Covid-19 containment measures, an endemic phase ensued, which may be traced back to DPCM dated May 4, 2020, whereby local circulation was reopened, and to DPCM dated 18 May, 2020, which allowed free circulation between regions. Such phase, marked on the map with a double black line, also concerns the entire Italian territory.

[ci] As specified above, these are Municipalities located near Codogno: Bertonico, Casalpusterlengo, Castelgerundo, Castiglione D'Adda, Fombio, Maleo, San Fiorano, Somaglia and Terranova dei Passerini.

In sum, reflexive cartography (Casti, 2015) makes it possible to envisage a space–time analysis of measures activated in the Italian territory to contain the Covid-19 epidemic in the months under study, i.e., between February and June 2020. Such cartography highlights: initial containment measures upon emergence of contagion, which affect the entire national territory; more stringent measures in the epidemic phase, which are gradually extended from an initial "red zone" area to the regions most affected and eventually to the entire national territory; and, finally, a progressive lifting of restrictions throughout the national territory as the infection starts to decrease and an endemic phase is reached. This type of cartography uses reflexive color-coding to provide an effective graphic visualization of evolving political measures over months: white marks the initial alert phase; various shades of gray refer to the period of maximum emergency, in relation the many territories affected; black marks a virtual return to normality as containment measures decrease.

Overall, reflexive cartography provides a bird's eye view of the measures adopted. In particular, it enables researchers to analyze the measures deployed in Italy to limit the Covid-19 epidemic in a space–time perspective, by cross-referencing them to the stages of contagion.

6.1.5 Conclusions

Analysis of the measures adopted makes it possible to convey the features of a wider framework of operable restrictions, definable in terms of the categories involved and arranged at multiple institutional and territorial levels. With an eye on the assessment made by the EU (EU, 2020), in an effort to draw up a roadmap for alleviating restrictions—which in fact may also be taken in terms of an actual tightening of measures—we can outline a few major areas of restriction:

– restrictions on the circulation of people and goods throughout the territory:

- closure/reopening (to be applied gradually) of EU internal and external borders for the movement of goods;
- closure/reopening (to be applied gradually) of EU internal and external borders for the movement of people;
- measures of restriction/reopening of national domestic circulation, between regions, entry/exit protocols for delimited areas (isolated controlled zones);

– restrictions on activities, broken down on the base of separate activity categories:

- schools for infants, children, teens, and adults; levels to be broken down according to varying degrees of learners' self-sufficiency in either classroom or online schooling;
- commercial activities, which may be broken down by size and format, and along different modes of aggregation in trade clusters or networks;
- social and hospitality activities, restaurants, bars, theaters, cinemas, hotels, sports centers, which may be broken down according to economic sustainability in the application of health safety measures;
- social, promotional, recreational activities connected to events (fairs, parties, major concerts, …).

The gradual reopening of economic and socioeducational activities will of necessity consider either the maintenance or subsequent lifting, in gradual steps, of social distancing measures and of health and safety rules in workplaces.

Analyses conducted on the dynamics of diffusion/intensity of contagion and on the countermeasures adopted in different contexts, must be related to distinctive territorial features (Walker et al., 2020). The demographic and social setup, the availability and quality of health care combine and substantially affect the impact of measures to contain viral spread. Given our as yet partial knowledge of contagion dynamics, it is arduous to assess the effectiveness of the range of measures which may be adopted as outlined above.

While our understanding remains for now incomplete, we may advance two main considerations with regard to the range of policies and decision-making tools authorities may decide to adopt. The first concerns a highly complex field of tension, generated by the search for the best possible trade-off between measures to restrict economic and social activities, which have (negative) effects on the well-being of individuals and the collective, and the concerns of public healthcare, which benefits from diminished contagion exposure. While it remains hard to discern a clear action scenario or to envision effective intervention, one second note worth considering has to do with the importance of timeliness for effective epidemic containment: "(…) In the absence of a vaccine, all governments are likely to face challenging decisions around intervention strategies for the foreseeable future. However, the still relevant counterfactual of a largely unmitigated pandemic clearly demonstrates the extent to which rapid, decisive, and collective action remains critical to save lives globally" (Walker et al. 2020, p. 422).

Along these lines, within a European context that—between February and June 2020—struggled to develop coordinated and homogeneous sets of political intervention, reflexive mapping serves to highlight the speed of intervention in Italy, where contagion containment policies are progressively modified as infection evolves: from an onset phase which recorded the first cases; to the epidemic stage which saw an exponential rise in contagion; and, finally, to the endemic phase as contagion is gradually contained. Ultimately, reflexive cartography is an effective tool for guiding a spatio-temporal analysis of the epidemic closely tied to the measures adopted by individual states. In particular, it highlights the pioneering role of Italy in the European context as a political reference model for emergency management.

References

Casti, E., 2015. Reflexive Cartography: A New Perspective on Mapping. Elsevier, Amsterdam-Waltham.

Casti, E., 2020. Geografia a 'vele spiegate': analisi territoriale e mapping riflessivo sul Covid-19 in Italia. Documenti Geografici, pp. 61–83.

Daniele, L., 1989. voce Circolazione delle merci nel diritto comunitario in Digesto delle discipline pubblicistiche. Utet, Tourin.

European Commission-Directorate-General for Communication, 2020. A European Roadmap to Lifting Coronavirus Containment Measures. https://ec.europa.eu/info/sites/info/files/factsheet-lifting-containment-measures_en. pdf.

European Council, 2020. Covid-19 Coronavirus Outbreak and the EU's Response. https://www.consilium.europa. eu/en/policies/coronavirus/.

Flaxman, S., et al., 2020a. Report 13: Estimating the number of infections and the impact of non-pharmaceutical interventions on Covid-19 in 11 European countries. In: Imperial College Covid-19 Response Team.

Lumet, S., Enaudeau, J., 2020. Organisation du territoire européen en temps de Covid-19, entre coopération etrepli. Le Grand Continent. https://legrandcontinent.eu/fr/2020/04/01/organisation-du-territoire-europeen-en-temps-de-covid-19-entre-cooperation-et-repli/.

Rossi Dal Pozzo, F., 2012. voce Circolazione e soggiorno nell'Unione europea in Digesto delle discipline pubblicistiche. Utet, Tourin.

Walker, P.G.T., et al., 2020. The impact of Covid-19 and strategies for mitigation and suppression in low- and middle-income countries. Science 369, 413–422.

Further reading

Allain-DuPré, A., et al., 2020. The Territorial Impact of Covid-19: Managing the Crisis Across Levels of Government. OECD. https://read.oecd-ilibrary.org/view/?ref=128_128287-5agkkojaaa&title=The-territorial-impact-of-covid-19-managing-the-crisis-across-levels-of-government.

Camera dei Deputati-Servizio Studi, 2020. Misure sull'emergenza coronavirus (Covid-19)–Quadro generale.

Chopin, T., Koenig, N., Maillard, S., 2020. The EU Facing the Coronavirus. A Political Urgency to Embody European Solidarity. Policy Paper, vol. 250 Jacques Delors Institute, Notre Europe, pp. 1–8.

Flaxman, S., et al., 2020b. Estimating the effects of non-pharmaceutical interventions on Covid-19 in Europe. Nature 584, 257–261. https://doi.org/10.1038/s41586-020-2405-7.

Gressani, G., 2020. Le Coronavirus à l'échelle régionale. Le Grand Continent. https://legrandcontinent.eu/fr/2020/03/17/le-coronavirus-a-lechelle-pertinente/.

Lapatinas, A., 2020. The Effect of Covid-19 Confinement Policies on Community Mobility Trends in the EU. EUR 30258 EN, Publications Office of the European Union, Luxembourg, https://doi.org/10.2760/875644.

Containment measures in relation to the trend of infection in Italy

Fulvio Adobati, Emanuele Comi, and Alessandra Ghisalberti*

6.2.1 Introduction

The range and number of measures issued to deal with the Covid-19 health emergency has given rise to an intricate scenario, which makes it hard to clearly isolate general, rational grounds underlying the guidelines set out in individual provisions (Camera dei Deputati-Servizio Studi, 2020). Analysis of data related to the results of public containment measures highlights such difficulty, as focus needs to alternate between an overview of swab-tests which returned positive outcomes to an assessment of the sanctions issued against citizens. Such hindrance has partly to do with Italy's institutional framework, namely the two levels of regulatory measures and administrative Acts of the national State and the Regions (Capano, 2020).

Specifically, the definition of "red zones," or areas subject to heavy restrictions (especially on people's mobility) was taken up in many provisions (see Fig. 6.5), issued in fairly rapid succession at specific epidemic stages (Danzi et al., 2020). After an "experimental" application of measures by the national government in the first days of the epidemic, the establishment of red zones was taken up in regional provisions that applied them a varied range of differently sized local areas. As result, while some regions repeatedly established "red zones," other regions saw no intervention in that respect (a policy found, in fact, to be quite unrelated to the incidence of infection). Along these lines, intervention on the handling of public transport occurred solely in the Lazio Region, which set its own maximum public transport limit to 50% of usual commute rates, in the form of an extension to restrictions on public transport use based on previously existing social distancing rules.

*The chapter is the result of joint research carried out by the three authors. More specifically, Fulvio Adobati compiled Sections 6.2.1 and 6.2.5; Emanuele Comi Sections 6.2.2 and 6.2.6; Alessandra Ghisalberti Sections 6.2.3 and 6.2.4.

Ultimately, an analysis of the measures affecting two key sectors of regional action, health protection and public transport services, brings out a complex chart of the provisions that were set in place. This in turn yields valuable evidence for assessing the shortcomings in the current framework of competences and in coordinated action at different territorial levels.

6.2.2 Main public measures for infection containment in Italy

As mentioned in the final section of Chapter 6.1, during the first wave of the Covid-19 epidemic that is the subject of our study, the Italian legal system intervened in timely and articulate fashion, providing for a series of unprecedented restrictions on individual freedoms.

While initial measures came under the wider government purview of Civil Protection,[a] legislators endeavored from the start to produce an extraordinary legislation *corpus*—consisting of decree laws adopted pursuant to art. 77, paragraph 2 of the Constitution—which set the framework whereby formally administrative Acts (i.e., the Decrees of the Presidency of the Council of Ministers or DPCM) laid down thorough rules on restrictions to individual freedoms. In other words, in order to safeguard the reserve of law guaranteed by the Constitution with regard to limitations on freedoms—decree laws would set out all the possible limitations which DPCMs could provide for as well as the legal limits of sanctions that could be imposed.

The first Decree of the Presidency of the Council of Ministers (DPCM) adopted on the basis of emergency legislation was dated 23 February 2020. It laid out the so-called "red zones" in the municipality of Vo' Euganeo in the region of Veneto and in the Codogno area in the region of Lombardy.[b] As mentioned earlier (Chapter 6.1) such Act imposed stringent restrictions on activities and movement, albeit exclusively for the red zones outlined at the time.

As of 1 March the area affected by restrictive prescriptions was extended to Northern Italy[c] with bans on activities that could lead to crowding, such as conferences, meetings or sports competitions open to the public. As of 4 March schools were closed. Finally, as of 8 and 9 March, with two DPCMs, travel was restricted unless justified by necessity or work reasons.

As of 11 March, all nonfood businesses or markets, even outdoors, which did not provide basic necessities were closed.[d] As of the same date, bars and restaurant businesses were also shut down, with the exception of home delivery.[e] Hairdressers and beauty parlors were also suspended.[f] All previous restrictions were equally extended to all of Italy.

[a] The Council of Ministers resolution of January 31, 2020 declaring a state of emergency was adopted on the basis of Legislative Decree no. 1, just as the reiterated declaration of a state of emergency was issued in order to ensure continued observance of Civil Protection rules.

[b] The municipalities included (a) Bertonico; (b) Casalpusterlengo; (c) Castelgerundo; (d) Castiglione D'Adda; (e) Codogno; (f) Fombio; (g) Maleo; (h) San Fiorano; (i) Somaglia; (j) Terranova dei Passerini.

[c] Namely: the regions of Emilia-Romagna, Lombardy and Veneto, as well as the provinces of Pesaro-Urbino and Savona.

[d] DPCM 11 March 2020, art. 1, no. 1.

[e] DPCM 11 March 2020, art. 1, no. 2.

[f] DPCM 11 March 2020, art. 1, no. 3.

A noticeable clampdown on movements occurred as of 22 March,[g] as citizens were forbidden to leave the municipality of residence—except for proven work reasons or urgent needs. Also, as of 23 March,[h] a general closure of production activities deemed nonessential was mandated.

Restrictive measures on productive activities and the prohibition to leave one's municipality ceased to be enforced as of May 4, 2020, while free mobility between regions was allowed to resume only June 3.

6.2.3 Results of containment measures on infection progress

The outcome of containment policies against the trend of positive Covid-19 cases may be assessed by monitoring the quantitative evolution of recorded infections, based on swab-testing. This is done by plotting the contagion index evolution based on the number of positive cases recorded in relation to swabs administered between 23 February and 30 June 2020 in Italy for every 1000 swabs, as shown in Fig. 6.3. Within this timeframe, we can zero in on changes in the positive case index 15 days after the four main DPCMs outlined above were issued[i]: DPCMs 23 February and 22 March 2020 during the "epidemic" phase; and DPCMs 4 May and 3 June during the "endemic" phase.

More precisely, if we focus on the "epidemic" phase—that is, the one in which the SARS-CoV-2 epidemic grew exponentially[j]—, the two main measures aimed at countering the spread of infection through unprecedented restrictions were DPCM of 23 February and DPCM 22 March 2020. As mentioned, in order to contain the health emergency, for the first time in Italy, the former imposed restrictions on individual freedom of movement and established the so-called "red zone" in the municipality of Vo' Euganeo in the Veneto region and in the Codogno area in Lombardy; the latter mandated a shutdown of all production-industrial activities considered nonessential on the entire national territory.

Fifteen days after the first DPCM was issued, namely as of March 9, the contagion index curve was still clearly on the surge: positive cases increased 150 on March 9 to 180 on March 22, for every 1000 swabs. This figure is well above the national average recorded during the entire period under study, equal to approximately 68 positive cases per 1000 swabs, which shows that containment policies implemented up to that date were falling short of desired positive outcomes.

---

[g]Ordinance of the Ministry of Health of 22 March 2020.

[h]Via a DPCM issued on the same date.

[i]For an analysis of the vulnerability of the Italian healthcare system, exposed during the SARS-CoV-2 epidemic in spring 2020 and, in particular, for a discussion of the fragmentation in regional policies or agencies that affected swab-testing, see Chapter 5.3 in this volume.

[j]This study adopts the timeframe model of SARS-CoV-2 infection, broken down by Emanuela Casti into the three stages or phases: "Onset phase," when initial cases are recorded; "Epidemic phase," when the virus spreads to the entire community; "Endemic phase," when the number of infections decreases quantitatively. See: Casti (2020), pp. 65–66.

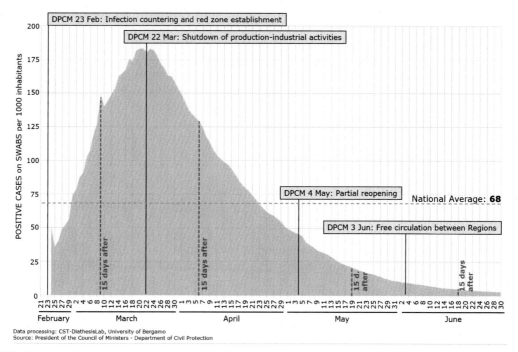

FIG. 6.3   Number of positive cases on swabs carried out in Italy, between 23 February and 30 June 2020.

Conversely, it may be noted that the index of positive cases on swab-tests carried out that far marked a downward trend 15 days after the second aforementioned epidemic phase DPCM was issued, as of April 6 to be exact. On that date, the index of positive cases reached levels below the national average and, at the end of April it reached a relatively contained level, equal to about 55 positive people for every 1000 swab-tests. We may thus conclude that the drop in positive cases as of that date could be ascribed to containment policies.

Once we turn to consider the "endemic" phase, when Covid-19 infection seems under control even though still present, the two main decrees that made it possible to relax restrictions and favor a return to normality in the people's movement, first between municipalities and then between Italian regions, were DPCM May 4 and DPCM June 3, 2020. As noted above, the former ordered a partial reopening of production and commercial activities, while the latter reestablished free circulation across the Italian territory. In this case, the evolution of the positive case index 15 days after both the first and the second DPCM—that is as of May 19 and June 18, 2020—follows a steady downward trend, down to a count of only 10 positive cases for every 1000 swab-tests (Fig. 6.3).

Overall, the evolution of the positive case index on swab-tests yields only a few useful criteria for evaluating the effectiveness of legal measures issued in Italy. In fact, numerical analysis for April 2020 records a downward trend of the curve for positive cases against administered swab-tests, due to the shutdown of all productive activity. The downward trend in infections is evident in the months of May and June 2020, which suggests that the Covid-19 epidemic across the Italian territory was under control. This favorable outcome may be ascribed to multiple concurrent factors, among which a persistence of lockdown measures and of containment policies adopted at national level.

6.2.4 Outcomes of containment measures on citizens fined by the police

An additional dataset which may help us monitor the outcomes of containment measures promoted in Italy concerns the quantitative evolution of the penalties or daily reports filed by the police, based on people checked across the national territory (Fig. 6.4). Such data attests to the level of citizens' compliance with containment measures adopted by the Government. It also records variations in inhabitant behavior based on the type of penalties which, in the Italian case, were turned from penal sanctions to administrative fines via legislative decree no. 19 of 25 March 2020. This second type involves an immediately enforceable monetary fine that citizens seemingly consider both more timely and more practical.

Trend analysis on the index of fined/reported individuals during checks presents fluctuations in the "epidemic" phase of the SARS-CoV-2 infection, i.e., starting from 11 March[k]

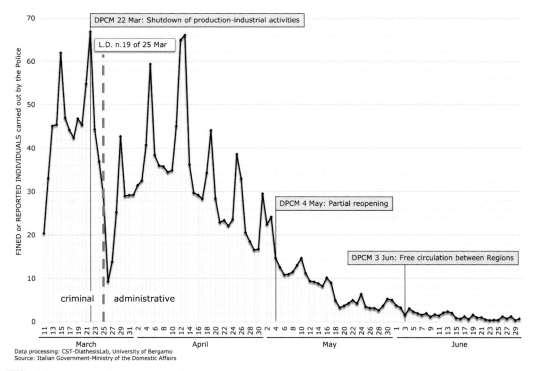

Data processing: CST-DiathesisLab, University of Bergamo
Source: Italian Government-Ministry of the Domestic Affairs

FIG. 6.4 Number of fined or reported individuals over 1000 checks in Italy (11 March–30 June 2020).

[k]Daily data on the monitoring of police activities for epidemic containment were published by the Ministry of the Interior starting from 11 March 2020 and are available at: https://www.interno.gov.it/it/coronavirus-i-dati-dei-servizi-controllo.

and up to DPCM 4 May 2020 which ordered a partial reopening of activities and mobility. An initial period may be noted, starting from the beginning of March 2020, as legislative decree 6 of 23 February 2020 came into effect, providing for criminal sanctions in the event of failures to comply with containment measures. A high number of fined/reported individuals was in fact recorded during checks. This period lasted until the end of March 2020. Successively, legislative decree 19 of 25 March 2020, which turned fines from criminal to administrative, marked a decreasing trend in the index of fined/reported individuals during checks. Despite fluctuations, this trend may be said to have lasted approximately until the end of April. Such favorable evolution is presumably ascribable to the pervasiveness of the administrative-pecuniary fines, which citizens seemingly regard as more immediate and more effective.

Finally, as the "endemic" phase of contagion was reached, and most notably following the DPCM May 4, 2020, which set out the progressive lifting of restrictions, data tend to settle and record a progressive decline in citizens' infractions. Actually, in May and June sanctions/complaints against the number of checks decreased until they reached nearly-zero levels. Obviously, these were the months which followed a relaxation of containment policies and also marked a return to normality in production activities and in the circulation of individuals. Progressively less restrictive policies were introduced, which consequently resulted in fewer violations of containment measures on the part of inhabitants and further lifting of restrictions.

Overall, the index of fines/reports during checks carried out by the police provides further, useful data for monitoring the outcomes of containment measures in Italy, and for recording variations in instances of irregular behavior on the part of citizens. While overall infractions were high in March and April 2020, they decreased significantly from May and reached minimum levels by June.

6.2.5 A difficult balance between Regions and the State: healthcare and transport

Albeit relatively recent,[1] Italian regionalism has undergone acceleration after the 2001 constitutional reform.[m]

As we turn to examine public policies of infection containment and look for hypotheses which, if backed by scientific data, may explain different contagion trends in neighboring areas, we may home in on two aspects whose organization depends on the Regions and which may have had an impact on the containment or the spread of the virus: healthcare organization and local public transport (Carullo and Provenzano, 2020a,b).

6.2.5.1 The health service

Ever since the historical, initial institution of regional administrations, tasks related to healthcare organization were transferred to the Regions.[n] Following the legal establishment

[1]Although already outlined by the Constitution, the Regions with ordinary statute began to exercise their autonomy only from 1970, having been established by Law 16 May 1970, no. 281.

[m]Enshrined in Constitutional Law no. 3/2001 which reformed Title V of the Constitution.

[n]Presidential Decree (DPR) no. 4/1972 and Presidential Decree (DPR) 9/1972.

of the national health service,[o] it was decided that such autonomy should be maintained, and key tasks in the provision of services were assigned to municipal administrations.[p] The State, or more precisely, the Ministry of Health was still charged with planning, supervision and coordination, as well as with tasks that called for unified enforcement throughout the national territory.[q]

Due to public finance demands and against the proven inefficiency of a centralist system of organization and management, a set of sweeping reforms was passed by Legislative Decree 502/1992 and Legislative Decree no. 517/1993. These introduced a management model based on corporatization and marked regionalization, which deprived both State and Municipalities of previous competences. In the ensuing overhaul, multiple functions were diverted onto Regions, for what concerned both the planning and coordination of facilities and the provision of healthcare services. The State was left with the sole tasks of financing what was by then a regionalized healthcare system and of defining essential healthcare levels in concert with Regions.

Regions were thereby granted considerable latitude in healthcare management, against possible conflicting competences from either State or Municipalities.[r]

As highlighted in the previous chapter,[s] regional choices were varied and distinct. In our assessment of data on contagion curve trends—especially in the ex-ante absence of healthcare or medical practices shared by the scientific community—different management choices made by the regions most involved in development may be taken as measures of the impact management policies had on contagion trends.

6.2.5.2 Local public transport

Local public transport is also managed by the Regions, even though, unlike healthcare competences, competences in the field of transport were only relatively recently assigned to Regions. In particular, starting with the Law of 15 March 1997, no. 59 (otherwise called as Bassanini Law)[t] a series of tasks and functions were entrusted to local administrations, among which were also tasks related to local public transport.[u]

The functions and tasks attributed to the Regions are quite extensive, and the State only holds prerogatives linked to international transport, road safety and pollution.[v]

Due to the emergency nature of legislation on Covid-19 containment, which was addressed in the opening chapters of this book, annex 15 of DPCM May 17, 2020 only mandated a set of

[o] Law 23 December 1978, no. 833.

[p] Art. 13 Law Number 833/1978.

[q] Art. 6 Law Number 833/1978.

[r] The Ministry does retain a monitoring prerogative, and competence in determining minimum healthcare still rests with the State and Regions Conference.

[s] See Chapter 5.3 in this volume, where the author discusses among other things data related to regional healthcare.

[t] Art. 4, paragraph 4 of Law 15 March 1997, no. 59.

[u] Legislative Decree 19 November 1997, no. 422.

[v] Art. 4 Legislative Decree no. 422/1997.

minimum prescriptions around contagion prevention: namely, the need to ensure a safe social distance of at least 1 m; the use of PPE (Personal Protective Equipment); and vehicle sanitization.

On May 22, 2020, the State and Regions Conference adopted guidelines which acknowledged DPCM recommendations, thus imposing a safe social distance of at least 1 m; the obligation to wear protective equipment on public transport; and the obligation to carry out periodic vehicle sanitization. Individual regions were left to independently determine safe capacity limits for mass transport.

Following the insurgence of coordination issues for local public transport and an increase in passenger traffic, the national government intervened by dictating uniform rules for all regions via DPCM 7 August 2020, whereby the need to ensure a 1-m social distance was waived to ensure a 60% maximum capacity for public transport vehicles.[w]

Finally, as the school year approached, in order to unify regional provisions and to address the issue of school transport, on 31 August 2020 the State and Regions Conference decided to extend maximum transport capacity to 80% of ordinary capacity.

Regional rules were partly required by the restart of activities once the restrictions were lifted and the school year resumed. By then the prospect of an increase in available means of transport had faded, even though derogations of ordinary rules had been granted.[x]

6.2.5.3 Administrative functions and the clash of competences: Regional measures on "red zones" and transport management

The peculiar allocation of competences between State and Regions, typifying Italy's constitutional setup after the 2001 reform, has been considered by some as a significant factor for analyzing contagion trends and assessing the effectiveness of SARS-CoV-2 containment strategies.

We would, however, be inclined to agree with what other observers have noted, also in view of what has emerged in the pages above: in the matter of healthcare organization and local public transport, the ordinary setup of administrative competences does recognize a regional authority. Nonetheless, there also exist, in situations of emergency, extraordinary, "monochrome" powers (*in broad terms* of health and safety) which pertain to both state and local authorities, equally competent to face the emergency.

Since ancient times, provisions for simplified and expedited procedures in emergency situations have been a common legal feature, bound to coexist—however—with ordinary competences to be exercised in ordinary situations. Such overlaps, while not invariably free from conflict, make it possible to ensure prompt intervention by the administrative authority, whatever it is. In such cases, necessity was deemed a legitimate source of power, overriding ordinary principles of autonomy, differentiation or adequacy.

[w] Annex 15, paragraph devoted to "the sector of local public transport either by car, or on lakes, lagoons, along coasts or railways not interconnected to the national network."

[x] Although, in some cases, even highway code exceptions had been provided for, in terms of vehicles deemed suitable for local public transport (for example, Article 200 of Legislative Decree no. 34 of 19 May 2020 provided that "notwithstanding Art. 87, paragraph 2, of the highway code, even cars for use by third parties referred to in article 82, paragraph 5, letter b, of the same code could be used on line services for the transport of persons") and art. 1 Legislative Decree July 16, 2020, no. 76 permitted simplified procedures for awarding local public transport services.

In the case of the Covid-19 epidemic, however, this dual level of competences (both for the ordinary administration and for the extraordinary powers of necessity and urgency) has revealed stretch marks in the legal fabric of Italian institutions. In particular, the legal system seems to have entered a critical period for many reasons, not all of which, in the subdued opinion of the present writer, are due to mere technicalities.

In the first place, given the absence of shared medical/scientific protocols, administrations were faced with a wide range of possible interventions whose effectiveness remains doubtful (think, for instance, of the prescription to wear masks or the policy of separating infected subjects from noninfected ones). Lack of technical knowledge amplified the scope of administrative discretion, calling upon administrations to make choices largely devoid of objective criteria or, in any case, of criteria shared by the medical community. The choices were, as a result, quintessentially political. Each decision-making level, therefore, acted as it saw fit. By way of example, if—on the contrary—we had been called to face a flood or an earthquake, intervention choices and methods would probably have been identical at any administrative level: aid would have been summoned, accommodation provided for displaced people and field hospitals set up to treat the wounded.

Also, the emergencies that the Public Administration was used to dealing with in the past were of a contingent and temporally limited nature: always think of the flood or the earthquake. Emergency situations that require the adoption of emergency administrative measures tend to be very limited in time and is aimed at regulating the first phases of emergency. Once the situation has stabilized, and even though administrative incompetence may still occur during the reconstruction phase (think of the "recent" earthquakes), competences tend to fall back into more sedimented patterns. In the case in question, the situation never quite stabilized, not even in terms of regulatory stability.

It should also be noted that measures for enhancing public services require operations which, in normal situations, require a long time to be implemented. Consider, for example, the purchase of goods and services which entails carrying out publicly accountable procedures. Although exceptions to such procedures were granted in order to speed up the purchase process, local administration offices—notoriously understaffed and little used to managing complex procedures—were ultimately unable to carry out the envisaged enhancement measures.

The response was therefore disjointed and fragmented, with local provisions at variance with national measures, delays in the implementation of policies and in the attainment of projects. In all likelihood, administrative uncertainty made it more difficult to cope with the emergency.

Specifically, as shown in Fig. 6.5, a response broken down in specific regional ordinances took place in spring 2020 for the establishment of "red zones," as well as for the management of public transport. The former is color-coded in red and refers to municipalities, within the Regions that established them, color-coded in pink; the second is marked by a green outline of the region which put in place specific management policies for public transport.

With regard to "red zones," we are presented with multiple regional ordinances which cordoned off as "red zones" two entire provinces of Emilia-Romagna—those of Piacenza and Rimini—as well as about 90 municipalities distributed in 12 different Italian regions.[y]

---

[y]Data on single regional ordinances, including their duration and the issuing body were collected by the Presidency of the Council of Ministers-Department of Civil Protection and are available at: https://github.com/pcm-dpc/COVID-19/tree/master/aree.

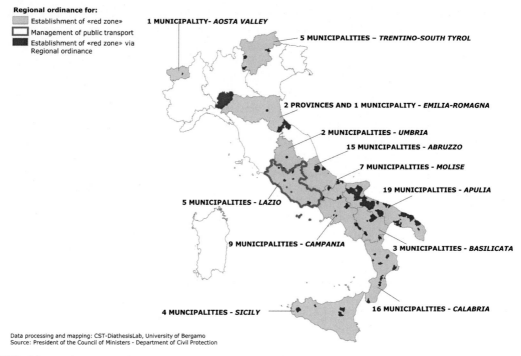

Regional ordinance for:

Establishment of «red zone»

Management of public transport

Establishment of «red zone» via Regional ordinance

1 MUNICIPALITY- *AOSTA VALLEY*

5 MUNICIPALITIES – *TRENTINO-SOUTH TYROL*

2 PROVINCES AND 1 MUNICIPALITY - *EMILIA-ROMAGNA*

2 MUNICIPALITIES - *UMBRIA*

15 MUNICIPALITIES - *ABRUZZO*

7 MUNICIPALITIES - *MOLISE*

19 MUNICIPALITIES - *APULIA*

5 MUNICIPALITIES - *LAZIO*

9 MUNICIPALITIES - *CAMPANIA*

3 MUNICIPALITIES - *BASILICATA*

4 MUNCIPALITIES - *SICILY*

16 MUNICIPALITIES - *CALABRIA*

Data processing and mapping: CST-DiathesisLab, University of Bergamo
Source: President of the Council of Ministers - Department of Civil Protection

FIG. 6.5   Establishment of "red zones" and management of transport via regional ordinances in spring 2020.

Such ordinances account for the wide fragmentation of the most restrictive measures across the national territory, attempts to integrate national policies which were slow to take action or proved far too limited, and thus inadequate to face the health emergency.[z] These betray a lack of health protection policies on a national scale, albeit in an extreme emergency situation, as well as a discrepancy between the levels of regional and state competence. This is a political fragility that may have negatively affected the management of the Covid-19 epidemic.

As regards the management of public transport, the Regions operated within the framework of the aforementioned national provisions and the Guidelines on public transport approved on May 22, 2020 by the Conference of Regions and Autonomous Provinces. Without prejudice to the provisions regarding the mandatory use of masks on board for all passengers and the application (with obvious limitations to control) of general health provisions (maximum body temperature of 37.5°, hand sanitation), the key issue revolved here around the progressive redefinition of maximum vehicle capacity. Maintaining a safe interpersonal distance of 1 m entailed a significant reduction in vehicle capacity, estimated at around 25%–30%, depending on the types of vehicles.

[z] These "red zones" were added to those established at national level by Decree of the Presidency of the Council of Ministers (DCPM) 23 February 2020 in the municipalities of Vo' Euganeo in the Veneto and the Codogno area in Lombardy; of 8 March 2020 for some regions of Northern Italy; and March 9 for the entire national territory, as discussed in Chapter 6.1 of the present volume.

In this sector, regional action saw interventions aimed at combining health safety conditions with transport needs tied to a projected demand for public transport upon reopenings scheduled for May 3, 2020. Along with policies to keep demand in check (especially, the widespread adoption of smart working), and to enhance supply by increasing public transport frequency, regional administrations issued specific provisions. In fact, via its own Ordinance dated 30 April 2020, only the Lazio Region—marked in Fig. 6.5 with the green outline—set a maximum capacity limit of 50%,[aa] which in fact was an extension to the previous public transport use limit which required safe social distancing. With respect to the public transport sector, therefore, the discrepancy between regional and state levels of competence would seem less evident, even though decision making in most Italian region was somewhat held back by a sense of "immobility."

6.2.6 Conclusions

In the form of cross-section of the many measures taken by the national government and regional administrations to combat contagion, our analysis marks an opportunity for reflection on a possible rearrangement of the competences involved in dealing with situations of emergency. The shortcomings outlined above call for an urgent assessment of what did work and of "areas of uncertainty" which paved the way to uncoordinated action and often led to controversial outcomes (Sanfelici, 2020).

The debate that has been raging for quite some time in Italy about a territorial overhaul and a review of national, regional and local competences, will be refueled by the shortcomings that the health emergency has highlighted. Proper attention will have to be paid to planning resilient models for territories exposed to emergencies of various kinds: the health emergency discussed above, but also, more generally, possible seismic or meteorological emergencies linked to climate change.

[aa] "Taking into account the obligation to use the masks, if it is not possible to continuously guarantee safe social distance within vehicles, pursuant to art. 3, paragraph 2, of the Decree of the President of the Council of Ministers of 26 April 2020, the service must in any case keep a maximum load not exceeding 50% of the transport capacity of the vehicle, as inferable from the vehicle registration certificate" (Ordinance of the President of the Lazio Region 30 April 2020, no. Z00037).

References

Camera dei Deputati-Servizio Studi, 2020. Misure sull'emergenza coronavirus (Covid-19)–Quadro generale.

Capano, G., 2020. Policy design and state capacity in the Covid-19 emergency in Italy: if you are not prepared for the (un)expected, you can be only what you already are. Policy Soc. 250, 1–8.

Carullo, G., Provenzano, P. (Eds.), 2020a. Le Regioni alla prova della pandemia da Covid-19, Dalla Fase 1 alla Fase 3. vol. I. Editoriale Scientifica, Milano (Abruzzo, Basilicata, Calabria, Campania, Emilia-Romagna, Friuli-Venezia Giulia, Lazio, Liguria, Lombardia, Marche).

Carullo, G., Provenzano, P. (Eds.), 2020b. Le Regioni alla prova della pandemia da Covid-19, Dalla Fase 1 alla Fase 3. vol. II. Editoriale Scientifica, Milan (Molise Piemonte Province Autonome di Trento e di Bolzano Puglia Sardegna Sicilia Toscana Umbria Valle d'Aosta Veneto).

Casti, E., 2020. Geografia a 'vele spiegate': analisi territoriale e mapping riflessivo sul Covid-19 in Italia. Documenti Geografici. 1, 61–83.

Danzi, V., Pinotti, G., Pisani, G., 2020. Provvedimenti regionali e emergenza Covid-19: un quadro generale. In: Carullo, G., Provenzano, P. (Eds.), Le regioni alla prova della pandemia da Covid-19. Dalla fase 1 alla fase 3. 2. Editoriale scientifica, Naples, pp. 739–757.

Sanfelici, M., 2020. The Italian response to the Covid-19 crisis: lessons learned and future direction in social development. Int. J. Commun. Soc. Develop. 2 (2), 191–210.

Further reading

Bertuzzo, E., et al., 2020. The geography of Covid-19 spread in Italy and implications for the relaxation of confinement measures. Nat. Commun. 11, 4264.

Casti, E., 2000. Reality as Representation: The Semiotics of Cartography and the Generation of Meaning. Bergamo University Press, Bergamo.

Casti, E., 2015. Reflexive Cartography: A New Perspective on Mapping. Elsevier, Amsterdam-Waltham.

Della Rossa, F., et al., 2020. A network model of Italy shows that intermittent regional strategies can alleviate the Covid-19 epidemic. Nat. Commun. 11, 5106.

Gentilini, U., Almenfi, M., Ian Orton, I., 2020. Social Protection and Jobs Responses to Covid-19: A Real-Time Review of Country Measures—A 'Living Paper'. Link https://www.undp.org/content/dam/south_africa/docs/Publications/global-review-of-social-protection-responses-to-COVID-19-2.pdf.

Migone, A.R., 2020. The influence of national policy characteristics on Covid-19 containment policies: a comparative analysis. Policy Des. Pract. 3 (3), 259–276.

Torri, E., et al., 2020. Italian public health response to the Covid-19 pandemic: case report from the field, insights and challenges for the department of prevention. Int. J. Environ. Res. Public Health 17 (10).

Weible, C.H., et al., 2020. Covid-19 and the policy sciences: initial reactions and perspectives. Policy Sci. 53, 225–241.

Conclusions: Towards spatial vulnerability management for a new "happy" living

Emanuela Casti

1 Territorial fragilities and containment interventions

A cursory assessment of the research indicates that the theoretical and methodological approach applied to the Italian case of the Covid-19 contagion has made it possible to track the viral spread on several scales in a space–time dimension. Data, capitalized and made available by regions, by provinces, and, in some cases, by municipalities via reflexive *mapping*, were cross-referenced diachronically to socio-territorial features. What emerged is that Italy's holistic features, both physical and territorial, have contributed to a marked epidemic differentiation. The distribution and regional evolution of the contagion in relation to population; the assumption of the Lombardy territory as a prototype of mobile and urbanized living; the cross-referencing of provincial and municipal data on population, mobility, pollution, hospital, and residential care facilities, and, finally, the specific focus on the area of Bergamo, all yield a composite picture, which certainly not only calls for further investigation but also provides solid initial grounds for interpretation.

Ultimately, the analysis conducted so far shows that the epidemic space is anisotropic and differs according to the fragilities of territories.

The first of these vulnerabilities lies in *pollution*, which intervenes in contagion either directly or by making our respiratory system prone to viral aggressions. This requires interventions to reduce the sources of polluting emissions (such as industry, livestock, domestic and industrial heating, and urban traffic among others): interventions that should be calibrated on the physical and social features of a specific region. We have found that because of a unique morphoclimatic setup, fine dust and polluting particles in the Po Valley stagnate in the air and accumulate over time, thus exceeding the safety thresholds and limits established by the European Union. Furthermore, the region is characterized by a high population density, which increases the magnitude of risk, endangering millions of individuals. Therefore, the first requirement for reducing epidemic damage is acting on air pollution. At the same time, such an action raises a more radical issue concerned with the climate crisis itself, and upholds the thesis of those who argue that epidemic crisis and environmental crisis are two sides of the same coin, and that defending ourselves from one also means facing the other (Horton, 2020; Lussault, 2020; Morin, 2020).

The second vulnerability is concerned with the *health and care system*, with regard to both the structural setup of health care and the ability to control infection and stem viral spread.

We have seen that epidemics require swift intervention and targeted actions aimed at equipping a territory with health-care facilities, which can act locally and systematically. In order to achieve that, it is essential to put in place appropriate facilities and to make sure institutional bodies respond in a coherent and timely manner to activate containment measures whenever a viral threat occurs. In Italy, where the national health system relies on policies that vary according to the regions, it emerged that highly centralized regions, with fewer and larger hospitals, were the most unprepared. In the case of Lombardy, for example, although health care is qualitatively excellent with regard to specialization, excessive centralization has deprived the region of a network of basic medical facilities, which would instead have provided first diagnostic and therapeutic aid independently of hospitals and prevented hospital overcrowding. What eventually emerged is the need for alternative buildings and containment spaces to hospitals, where asymptomatic or recovering individuals who must self-isolate from their cohabiting family members may be adequately assisted.

In the same manner, it became clear that the welfare system, and therefore long-term health facilities and retirement homes, must be protected from epidemics via adequate precautions, since they are places that may easily implode in case of viral outbreaks. The concentration of large-scale Nursing and Residential Care Facilities (RSAs) and the presence of many care homes in the north of the country contributed to a considerable surge in deaths, on account of both the advanced ages of the hospitalized guests and the management issues around these facilities. In fact, these facilities often employ precarious medical and nursing staff who work on a rotational basis in several hospitals, and who, in the event of an asymptomatic infection, unwittingly favor the spread of the virus.

In Italy, the uneven distribution of these hospitals, which varies regionally, must be assessed not only from a quantitative point of view but also on the basis of different care models provided to the elderly. This variability certainly depends on regional policies around the welfare infrastructure, but above all is tied to the range of cultural values present in the communities and on the organization of labor: the degree of female employment and the consideration in which the elderly are held within the community may be seen as the causes of the differences recorded. While in the north, the rate of female employment is high and recourse to hospitalization is virtually compulsory, in central and southern nonurban Italy, assistance for the elderly is mainly provided by the family, since female employment is more limited and high social value is attributed to the elderly. This makes home care more practicable, and in the event of an epidemic, it is the first line of defense for the elderly.

The third and final issue is concerned with some aspects *of contemporary living*, which are spatially expressed via *mobility and population density*. These affect the intensity of the contagion, since in times of an epidemic, "fragilities" arise within them, that is, space–time conditions—such as crowding and forced contacts between people on public — which favor viral spread. However, research results show that these aspects become "fragilities" only if population density and mobility are combined and if risk situations arise within them. As far as population density is concerned, cities should not be considered as risk areas in an absolute sense. One should rather consider public spaces and *hyperplaces* (such as stations, airports, shopping centers, etc.) that promote crowding and make it difficult to maintain social distancing. Similarly, within mobility, fragility lies in commutes carried out on collective public transport. Commuters forced to live in closed environments for even prolonged periods create ideal conditions for rapid viral transmission.[a]

In short, the Covid-19 epidemic in Italy has brought to the surface some risks of contemporary living and has forcefully brought to the fore the realization that the same reticular model of globalization must be managed and programmed so as to avoid them, especially in particularly dynamic and internationalized regions, such as those of Northern Italy. At the same time, however, the pandemic has shown that despite serious disruptions, complex societies such as Italy's remain largely able to cope with the crisis and provide for primary living needs: basic necessities and all essential services are guaranteed.

It should also be remembered that the current experience has considerably raised the awareness of the importance of communication technologies: remote connections and *smart working* have made up for reduced sociability and have partly addressed the need for interpersonal contacts. Home delivery services set up by shops and restaurants have pointed to different ways of experiencing private living spaces. As a result, inhabitants have experimented with new forms of sociality, which, when not imposed by the pandemic, may point to valuable alternatives for living a richer and more flexible life.

2 Territorial regeneration and potentials set forth by the pandemic

The range of territorial interventions to be undertaken, urgently in order to redress fragilities and stem contagion, is aimed at social distancing and the reduction of risk factors. Pending vaccination, the use of individual devices combined with the rapid track and tracing of infected individuals and subsequent self-isolation are the fundamental and primary line of defense against viruses. For this reason, it is imperative that the fragilities just highlighted be urgently addressed.

The identification of commuting as a cause of contagion, given that it produces gatherings and consequently spreads the virus not only by proximity but also by reticularity, opens up a more general reflection on mobility management. Mobility may be reduced thanks to technologies, which, precisely in this pandemic period, have shown their potential as exemption systems and means to replace meetings, contacts, training, and work activities. This potential must be managed to allow students and workers to stay at home or use coworking spaces near home, thus cutting down on commutes. In turn, commuting may be scheduled more efficiently: for example, by staggering shop and school opening times, promoting eco-friendly means of transport, and car sharing, which may ease pressure on collective means of transport. Interventions of this kind must be complemented by smart working, which ought to be made more widespread, since it is the most effective way of reducing commutes. Moreover, smart working cuts down air pollution: urbanized areas may be made healthier by combining smart working with energy-efficient models of housing, mobility, and transport.

[a]Since commuting is the most easily manageable form of mobility, one wonders why measures to tackle such issues have been half-hearted if not altogether ineffective in Italy.

Smart working provides a viable option that would have been unthinkable before the pandemic, namely, the choice of a place of residence detached from the workplace. If people were required to be physically present at their workplace only occasionally, they may opt to reside on the margins of cities, or farther inland, in mountains or sparsely urbanized areas, which present more favorable environmental conditions. Moreover, reduced commute times would mean individuals may devote more time to sports, children, hobbies, or other activities, which would greatly improve their life quality.[b]

With regard to interior living spaces, the epidemic has made it clear that our homes, especially condominium-style housing, require larger private spaces for smart working and common open-air spaces (balconies, gardens) that make up for the unavailability of public areas in times of a contagion.

Finally, the Covid-19 pandemic has highlighted infrastructure shortcomings in areas with a low urbanity rate: these must not only be considered but also properly addressed, in order to promote new ways of inhabiting territory that are safer, more balanced, more environmentally conscious, and more readily attuned to the needs of all generational components of the population (young people, adults, and the elderly).

In turn, all such interventions have promoted broader reflection on our model of inhabiting the earth.

3 The anthropocene era: Environmental and pandemic crises

The pandemic is forcing us to rethink our way of life with a renewed awareness (Ellis, 2020; Lucchi, 2016; Stiegler, 2019). In an era of "polychrysis" (Morin, 2020) that takes on the characteristics of a "syndemia", given that it manifests itself against a background of social and economic disparities that exacerbate the negative effects of every single crisis (Horton, 2020), the fraught relationship between humanity and the rest of the planet is becoming increasingly evident.

In short, the environmental crisis, which characterizes the Anthropocene as an era where earth features are strongly affected by human action and whose first systemic consequences bring about a "redistribution of the cards" of life on the planet (Crutzen and Stoermer, 2000), is today accompanied by recurrent health crises, which call for an all-out reflection on the model of inhabiting the earth.

Our role in the world must be rethought, overcoming the separation between culture and nature (Latour, 2020), in which humanity plans and implements a transition toward sustainable development, reactivating the ancient alliance between human and natural components. All inhabitants of the planet are responsible for redefining their relationship with natural resources with a view to patrimonializing its values. Quality of life depends on the satisfaction of global needs, such as pollution, fight against land consumption, and a balanced use of local resources, which may protect the environment and treasure landscape. To these global needs

---

[b] An urban exodus to the mountains has already started spontaneously in Italy. Holiday homes are being occupied permanently, and public tenders over abandoned housing have seen entire family groups moving inland.

are added those generated by the Covid-19 pandemic, which brings out the risks of contemporary living, on which this volume has focused.[c]

Among theories that provide analytical tools for addressing this challenge are those of complexity and reticularity. The former model states that a complex territory bears within itself autopoietic potential, which enables it to cope with crises, regain balance, and achieve a degree of complexity higher than the previous one (Raffestin, 1982; Turco, 1988). After a crisis, such as the epidemic one, and after an initial phase of imbalance, the territory becomes more stable, since it is equipped with defense mechanisms, which enhance its ability to cope with subsequent disturbances. This type of resilience should not be understood as a balancing response on the part of the environment. Rather, it should be seen as a driving force that makes the environment more complex: experientially richer and better equipped to face outside challenges. As it emerges from the crisis, a complex system aims at relaunching rather than at restoring its preexisting setup.

We gain a fuller picture of complexity theory, which diachronically outlines the evolution and potential of a complex territory, is complemented by a theory that aims to come to terms with reticular functioning of the contemporary world: a reticularity model we mentioned above, which integrates local territories with global networks. Thanks to globalization, the dynamics induced by the continuous flow of people and information amplify, but at the same time standardize the functioning of places, bringing out their integrative potential. As we have argued repeatedly in the pages of this book, this potential rests on the fact that individuals express the same model of urbanity, based on connected and widespread living, which reduces the distance and difference between center and periphery.

In Italy, this model of living has recently been assumed and envisioned in the *metro-mountain* formula: a living pattern that is no longer dichotomous or polarized around the densely urbanized areas of the plains, but is spread out to mountain areas in a complementary and dialectical relationship (Dematteis, 2018). We set out to implement a new metro-mountain and metro-rural vision, based on trailblazing models of interdependence and cooperation between different territorial systems, which may promote and develop a rhizome-like model of reticularity. Marginal areas offer a chance to rethink the habitation of places in innovative ways, starting with their specific cultural and environmental potential: marginal areas become a laboratory for testing new modes of living in times of crises (environmental and pandemic).

A metro-mountain vision may be tested precisely by combining the quality of environment and landscape with efficient citizen services: a model of living capable of offering at the same time greater environmental opportunities than living in the city and urban services for residing in the valleys and in marginal areas, by carrying out activities related to higher education or remote work.[d]

---

[c]See https://medium.com/cst-diathesislab; https://legrandcontinent.eu/fr/observatoire-coronavirus/.

[d]At the same time, this creates conditions for planning a new type of tourism, which may experience mountains in their many seasonal facets, thus promoting sports and a properly paced lifestyle attuned with nature.

By combining environmental balance and territorial development and by relying on the reticularity of relationships and the proximity of mobility for inhabitants, we can envisage a new living model based on the values and principles that increase life quality and set the conditions for a happy life, a subject we will address shortly. All this anticipates a season of welfare policies that may endow marginal and nonurban areas with a new shape, while at the same time ensuring that urban areas no longer act to centralize services but rather to integrate them.

As it moves from a cultural perspective that turns to earth's habitability in order to overcome the dichotomy between man and nature, the metro-mountain model ultimately envisions a transition toward sustainable development focused on infrastructural processes of marginal areas, and, at the same time, addresses highly urbanized areas from an ecological and environmentalist perspective.

Of course, the idea did not originate in Italy at this time. As early as 2013, the *National Strategy for Internal Areas* (SNAI)[e] set out to stem the depopulation of marginal areas, recognizing in these territories key opportunities for economic growth, which now must be updated to take into account the outcomes of the Covid-19 epidemic.[f] The great environmental and cultural resources of inland areas as well as their considerable and qualitatively relevant historical heritage—which lies mostly unused—may be effectively tied to the artistic and cultural heritage of urbanized areas, so as to yield enormous potential for a renewed idea of integrated development in the sign of environmentalism and technological infrastructure.

If we bring into sharper focus the relationship between inland areas of the peninsula and urban areas, we can assess how the epidemic has differently affected the territories. In inland areas, the epidemic acted as a detector and an accelerator of structural problems and social inequalities; a process that exacerbated, on one hand, historical gaps related to macrogeography, while also setting up conditions for a new peripherality. Urban areas, on the other hand, feature social inequalities and lifestyles that are remarkably costly in terms of environment, ecology, sustainability, and society. Accordingly, in times of an epidemic, urban areas require equally costly social-distancing measures.

The fact is that we now turn to the metro-mountain model as a perspective, whereby exchanges between inland areas (mountain, rural, and intermediate territories) and urbanized areas are seen with a renewed awareness. That leads to envisioning a territory that is no longer metrocentric, that is, centered around large cities, or even polycentric, that is, dependent on the medium-sized cities of the plains, but instead focuses on dynamism and mobility, in which material and digital resources intersect. In turn, that promotes a way of living places

---

[e]See the website https://www.agenziacoesione.gov.it/strategia-nazionale-aree-interne/.

[f]The National Recovery and Resilience Plan (PNRR) is an investment program that Italy must present to the European Commission as part of Next Generation EU, a tool to respond to the pandemic crisis caused by Covid-19. SNAI is thereby given a new key role, since the program entails ample funding and involves the transformation of the Internal Area National Strategy from an experimental policy over select pilot areas to a structural policy endorsed by the state.

that relies on a "rhizome-like" model, based on the multidirectional relationships determined by the daily choices of individuals.[g]

Public bodies need to translate the issues outlined above into concrete proposals for a bottom-up model of subsidiarity via institutions that are closest to citizens, as such institutions are better equipped to convey their needs and demands. The process should be implemented not only hierarchically, via various levels of government, but also horizontally, by favoring active involvement of inhabitants.[h]

Such complex cultural challenges require a participatory model that relies on interdisciplinary criteria in order to establish governance aimed at integrating local sets of knowledge and at activating processes that aim to cocreate knowledge and codesign new forms of spatialized welfare via integrated tools and practices. We envisage a form of participation based on a *Contract of Living*, an operational tool aimed to support municipalities and supramunicipal local authorities in building visions, strategies, and urban and architectural projects in a perspective of territorial networks able to convey a functional integration between *center and margin*.[i]

The *Contract of Living* should pertain to the entire conception–planning–execution process of intervention based on some assumptions. The first is that a territory is an environment instantly produced by all its inhabitants, either ephemeral or permanent, either power holders or ordinary citizens at all scales. It is precisely on this perspective that the *Contract of Living* arises as the outcome of political intermediation between administrators and citizens, which from the start involves inhabitants in innovative decision making as experts of their own territory able to interact with multiple institutional and noninstitutional actors.

Consultation must be implemented via performative communication tools, such as cybercartography, which can not only spatialize and bring into focus the social issues tied

[g] Operationally, we aim to actively strengthen systems and infrastructural networks (both physical and virtual), mobility networks (fast and slow mobility, sustainable transport, and local public individual transport), and environmental networks via public network services linked to water cycles, to the use of waste for producing energy, and to smart technological networks, but above all to welfare: spatial forms and organization meant to boost health care, training, and intergenerational sociality. Not only disused real estate assets but also techno-rural economic infrastructuring, such as agroforestry economies, ecosystem services, and innovative forms of agriculture, point to an unexpressed potential, which could effectively be tied to the system. Finally, infrastructuring of the green and blue networks linked to *climate change*, to the water systems and networks, to ecological corridors, and to territorial networks of hydrographic and river basins exhibit Italy's key feature: a country of many diversities and riches condensed within a limited national surface.

[h] The local institutional layout, comprising municipalities, mountain communities, and rain-collecting mountain basins play a crucial role in that respect. These bodies are in fact parties and contractors of an implicit *Contract of Living*, and are called to operate according to the canons of administrative law, in the constant search for balance between public interest and private freedom, aiming to ensure the full realization of people, quality of life, and, therefore, the achievement of well-being and happiness.

[i] Rejecting the dated top-down notion of a *master plan*, the *Contract of Living* brings to the fore inhabitants, that is, stakeholders and institutional interlocutors involved in shared urban-territorial planning: builders ready to cope with challenges and willing to identify and promote regeneration projects. Their involvement follows precise governance and participation guidelines and provides for compensation/reimbursement tools in the event the contract is breached (Lévy et al., 2018).

to territorial fragilities but also can promote its features and opportunities, so as to produce an empowering awareness of the advantages afforded by reticular regeneration. It is also with a view on all this, therefore, that we envisage a new mode of inhabiting territory: a mode able to tackle epidemics effectively while at the same time providing substantive data for addressing the challenges of the Anthropocene.

4 Happy living

An acknowledgment of the possibility of inhabiting territory "on a human scale" involves the concept of substantial equality. The notion is enshrined in Article 3 of the Italian Constitution, which dictates the removal of economic and social obstacles that may hinder the full development of a human being and advocates effective participation of all workers in the political, economic, and social organization of the country, with a view to achieving shared well-being.

Thus, we touch upon the cluster of values that underlie our view of a new mode of living and inhabiting, which may be condensed as follows: *the time has come to rethink our future not in terms of quantitative growth but in terms of life quality, to be achieved by reclaiming values that we have lost in everyday life.* It goes without saying that these include aesthetic values, such as the beauty of landscape, as well as ethical values, such as those inscribed in nature. Happiness, understood as a right to be exercised on an individual and social level, depends on such values.

The Fathers of the American Declaration of Independence turned to nature as a source of ethical and moral inspiration and as a basis for affirming the right of citizens to seek happiness. The UN, in turn, has recently endeavored to assess the quality of social life via happiness indicators, which are monitored and published yearly.[j]

Article 44 of the Italian Constitution, specifically aimed at the mountain context, also offers a basis for reflecting on a new model of living, which may reclaim the central role of nature to affirm the right to happiness. In fact, seen in territorial terms, happiness may be expressed as the right to welfare or as the so-called "social state," built and conceived as a collective value and as a result of an organized civil and political community (Ferrara, 2010). On the basis of this right, the happiness of the individual and that of the individual as a citizen may not be separated from the common good, or from a social action aimed at public happiness. On the contrary, happiness derives precisely from harmony between the pursuit of happiness as an individual aspiration and the concept of public happiness, attributable to the political community ruled by law.

It is in this perspective that the framework of the "system of values" just mentioned, which may be classified under Social Justice, Citizenship Law, and Sustainability, can be translated into the operational actions mentioned above, which are: in the political sphere, the overcoming of representative democracy through a *Contract of Living* (Lévy, 2019); in welfare, through a polycentric system of public health services of education and accessibility to be considered as common goods capable of guaranteeing inclusion of marginal territories; and finally,

[j]See United Nations, 2020.

in sustainability, via actions that protect nature, symbolically shown by the quality of landscape intended as an identity feature with a strong aesthetic value that ensures social reproduction.

Ultimately, the longing that the pandemic crisis has brought to the surface is not to curl up in self-defense, but to reach out and act to build a new, fine, and happy world.

References

Crutzen, P.J., Stoermer, E.F., 2000. The "Anthropocene". Global Change Newslett. 41, 17–18.

Dematteis, G., 2018. La metro-montagna di fronte alle sfide globali. Riflessioni a partire dal caso di Torino. J. Alpine Res. 106-2. https://doi.org/10.4000/rga.4318.

Ellis, E.C., 2020. Antropocene. Esiste un futuro per la terra dell'uomo? Florence, Giunti.

Ferrara, R., 2010. Il diritto alla felicità e il diritto amministrativo. In: Diritto e processo amministrativo, vol. 4, pp. 1043–1089.

Horton, R., 2020. Covid-19. La catastrofe. Il Pensiero Scientifico Editore, Rome.

Latour, B., 2020. La sfida di Gaia: il nuovo regime climatico. Meltemi, Sesto San Giovanni.

Lévy, J., 2019. Démocratie interactive: pour un grand débat. Fondation Jean Jaurès. accessed from January 2021 from https://jean-jaures.org/nos-productions/democratie-interactive-pour-un-grand-debat.

Lévy, J., Fauchille, J.N., Pòvoas, A., 2018. Théorie de la justice spatiale. Geographies du juste et de l'Injuste. Odile Jacob, Paris.

Lucchi, G.C., 2016. L'antropocene e il salto quantico. Tradizione ed evoluzione, Tipheret, Acireale.

Lussault, M., 2020. Chroniques de géo' virale. Ecole urbaine de Lyon/Editions deux-cent-cinq, Lione.

Morin, E., 2020. Sur la crise. Flammarion, Paris.

Raffestin, C., 1982. Potere e territorialità. In: Raffestin, C., Bruneau, M. (Eds.), Geografia politica: teorie per un progetto sociale. Unicopli, Milan, pp. 63–70.

Stiegler, B., 2019. La società automatica. vol. 1 L'avvenire del lavoro. Meltemi, Sesto San Giovanni.

Turco, A., 1988. Verso una teoria geografica della complessità. Unicopli, Milan.

United Nations, 2020. World Happiness Report 2020. accessed January 2021 from https://worldhappiness.report/.

Index

Note: Page numbers followed by *"f"* indicate figures.

A

Algorithm-driven processing, of big data, 3
Anamorphic map, 43, 84
Atmospheric pollutants, 24–25
  background environmental condition, 114
  basemap, 122
  carrier effect, 114, 123
  color coding, 119
  domestic heating, 115
  ESA, 115–116
  ISPRA, 117, 119
  nitrogen dioxide (NO_2), 113, 116, 118*f*, 124
  NW-SE Milan-Lodi-Cremona direction, 113, 124
  particulate matter ($PM_{2.5}$), 119–122, 124
  particulate matter (PM_{10}), 115, 119–122, 124
  peri-urban clusters, 117–119
  propagation dynamics, of contagion, 114
  space, 114, 124
  territorial Covid-19 contagion severity, 113
  territorial fragility, 114
  transport-related emissions, 115
"Azienda Zero,", 162–163

B

Basemap, 92–93
Bassanini Law, 211
Bergamo, 1, 52–56
Brembana Valley, 54–55

C

Case mortality rate (CFR), 66
Chorographic metrics, 12–13
Color-coded map, 83–84
Color-coding, 41–43, 106–107, 188–189
Consultation, 223–224
COMBO algorithm, 107
Commuting
  Covid-19 infection
    in Italy, 95–96
    mobility, European context of, 92–95
    proximity and reticularity, 91–92
  definition, 101
  in Lombardy, 99–103
"Contagion backbone,", 104–105

C (continued)

Containment measures, of Covid-19 infection
  administrative functions, competency clashes, 212–215
  health service, 210–211
  local public transport, 211–212
  outcomes, police penalties, 209–210
  public measures, for infection containment, 206–207
  "red zones,", 205, 213–214, 214*f*
  results, infection progress, 207–209
  social distancing, 205, 211–212, 215
  swab-tests, 205, 208
Contract of Living, 223–225
Cybercartography, 11, 223–224

D

Decrees of the Presidency of the Council of Ministers (DPCM), 172, 206–207
Digital mapping, 38, 189–190

E

Elderly residential hospitality, 169–172
Emergency administrative measures, 213
Environmental crisis, 220–224
"Epidemiological transition,", 147–148
Essential Levels of Assistance (LEA), 155–156
European Commission, 97
Europe, Covid-19 epidemic, 19, 20*f*
  color-coding, 22–24
  economic and social damage, 20
  extrusion, 22–24
  factors, viral propagation, 24–27
  geo-visualization, 22
  infographic processing, 21–22
  map's social implications, 21–22
  mosaic map, 20–21
  reflexive cartography, 21–22
  regulatory provisions and social outcomes, 20–21
  spatiality, 21–22
  statistics-based data processing, 21–22
  temporal frequency, 20–21
  territorial factors, 27
  tridimensional representation, 21–22
  viral incidence, 22–23
Eurostat indicators, 92

H

Heuristic map, 106–107
Home services, 146–147

I

Imagna Valley, 54–55
"Immobility,", 215
Infection speed rate, 3–4
Infection rate (IR), 66
Influenza-Like Illness (ILI), 157
Informal care, 148
Integrated Environmental Authorization (AIA), 122
International Long-Term Care Policy Network,
    144–145
ISPRA, 99
Italian National Institute of Statistics (ISTAT), 65–66,
    79–80, 103–105
Italy, Covid-19 contagion
    Bergamo, 52–56
    cartographic models, 51–52
    cartographic representation, 56
    color-coding, 55, 57
    containment measures implementation
        administrative functions, competency clashes,
            212–215
        health service, 210–211
        local public transport, 211–212
        outcomes, police penalties, 209–210
        public measures, for infection containment, 206–207
        "red zones,", 205, 213–214, 214f
        results, infection progress, 207–209
        social distancing, 205, 211–212, 215
        swab-tests, 205, 208
    epicenter regions, of Covid-19 epidemic, 161–165
    health care system and swab testing, 155–160
    mortality rates
        by age analysis, 74–76
        age-patterns, cross-country comparison of, 67
        counterfactual analysis, 66
        death rate, in March 2020, 68–70
        estimation, 70–74
        factors, 67
        ISTAT, 65–66
        population death rate, 67
        swab tests, 65–66
    nursing and residential care facilities (see Nursing and
        residential care facilities, in Italy)
    public policies (see Public policies for epidemic
        containment, in Italy)
    school mobility, 57–59
    Seriana Valley hotspot, 56–59
    socio-territorial aspects, 51
    surface extrusion, 55

L

Local Labor Systems (SLL), 104–105, 105f
Lockdown period, 97, 109, 189
Lombardy, 1
    high-crowding public spaces/hubs, 101
    infection intensity in
        communicative levels, changes of, 43
        connotational model, 43
        contagion index, 46
        data visualization, 46–47
        deflagration symbol, 46–47
        distortion techniques, 43
        distribution and temporal evolution of, 41–48
        epidemic propagation, 46
        extrusion, 46–47
        first epidemic wave, 41
        geolocalization, 46–47
        peripheral areas, 45–46
        reflexive cartography, 46–47
        self-referential information, 46–47
        socio-territorial factors, 41
        topographic metrics, 43
    interconnected living patterns, 100
    long-care system in
        accreditation mechanism, 175
        anamorphic distortion, 176–177
        elderly residential hospitality, 169–172
        heterogeneity, of RSAs, 175
        "multiple comorbidity,", 170–171
        non-self-sufficient people, national and Lombard
            care-system for, 172–175
        nursing and residential care facilities (see Nursing
            and residential care facilities)
        social welfare model, 169–170
        "specialized clusters,", 170
    mobility model, 101
    monitoring commutes in, 103–106
    mortality in
        age analysis, 85–87
        data analysis, 80–82
        estimation, 82–85
        outbreak severity, 79
    rhizome-like form of commuting in, 106–109, 107f
    urbanity and commuting, for reflexive mapping,
        99–103
Long-care system, in Lombardy
    accreditation mechanism, 175
    anamorphic distortion, 176–177
    elderly residential hospitality, 169–172
    heterogeneity, of RSAs, 175
    "multiple comorbidity,", 170–171
    non-self-sufficient people, national and Lombard
        care-system for, 172–175